HVAC CONTROL SYSTEMS NETWORKING

Basic Configurations and Troubleshooting

Second Edition

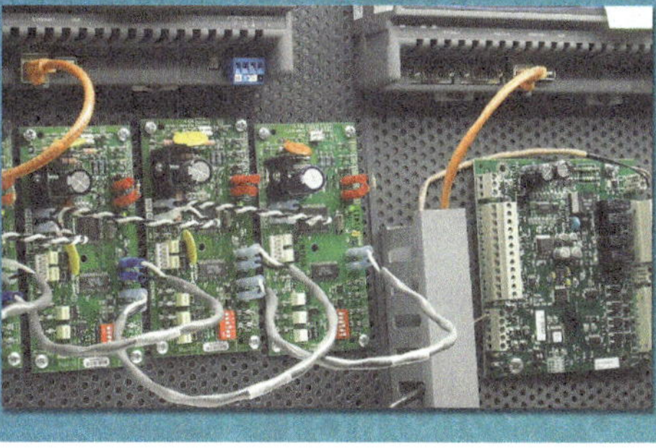

AMERICAN TECHNICAL PUBLISHERS
ORLAND PARK, ILLINOIS

Ronnie J. Auvil

HVAC Control Systems Networking: Basic Configurations and Troubleshooting contains procedures commonly practiced in industry and the trade. Specific procedures vary with each task and must be performed by a qualified person. For maximum safety, always refer to specific manufacturer recommendations, insurance regulations, specific job site and plant procedures, applicable federal, state, and local regulations, and any authority having jurisdiction. The material contained herein is intended to be an educational resource for the user. American Technical Publishers assumes no responsibility or liability in connection with this material or its use by any individual or organization.

American Technical Publishers Editorial Staff

Editor in Chief:
Peter A. Zurlis
Assistant Production Manager:
Nicole D. Burian
Art Supervisor:
Sarah E. Kaducak
Digital Media Manager:
Adam T. Schuldt
Technical Editor:
Julie M. Welch
Copy Editor:
Dane K. Hamann

Cover Design:
Bethany J. Fisher
Illustration/Layout:
Bethany J. Fisher

Photo Acknowledgment:
Fluke Corporation

2 3 4 5 6 7 8 9 – 19 – 9 8 7 6 5 4

Printed in the United States of America

ISBN 978-0-8269-0790-5
eISBN 978-0-8269-9535-3

This book is printed on recycled paper.

Contents

1 Networking Personnel 1

Networking Personnel

2 Networking Overview 5

Networking History • Networking • OSI Model • Ethernet Networks

3 HVAC Control System Networks 25

HVAC Control Networks • Working with Web-Based Supervisory Controllers

4 Client Computers 33

HVAC Control System Client Computers • Computer Software • Network Access • System Security • Controls Software

5 Servers 43

HVAC Control System Servers • Server Software • Server Setup and Maintenance • Basic Server Troubleshooting

6 Wireless Networking 55

Wireless Networking • Wireless Network Troubleshooting

7 Diagnostic and Troubleshooting Tools 63

Hardware Tools • Software Tools

8 Troubleshooting Scenarios 73

Troubleshooting • Common Scenarios

Appendix 89

Glossary 95

Index 99

Introduction

HVAC Control Systems Networking: Basic Configurations and Troubleshooting is an introduction to Ethernet networking concepts, hardware, configurations, and troubleshooting for non-IT professionals. This resource is essential for building operators and HVAC controls technicians working with web-based supervisory controllers.

This textbook covers the following:

- The roles and responsibilities of building operators, HVAC controls technicians, and IT staff
- General and Ethernet-specific networking fundamentals
- HVAC control system integration with computer networks
- Common requirements and configurations for client computers and servers
- Wireless networking
- Basic diagnostic and troubleshooting tools
- Descriptions of the simple problems that cause most network disruptions, along with recommended solutions

Free App Available

Available separately, this interactive mobile app includes the textbook troubleshooting scenarios in an easy-to-use field reference. A simple flow chart sequence is used to narrow down a problem to the most likely cause. Each result then describes the solution, along with advice on how to avoid the problem in the future. Also included in the app are definitions of common terms.

Look for *HVAC Network Troubleshooting* on the Apple® App Store® and Google Play®

Download on the App Store

GET IT ON Google Play

Networking Personnel

1

Most modern building systems, including HVAC control systems, utilize computers and network technology to perform sophisticated functions. Not only must a building operator or an HVAC controls technician be able to work on these complex systems, but they must also be able to collaborate effectively with a facility's IT staff.

NETWORKING PERSONNEL

Networking and IT technologies are becoming an ever-larger part of the HVAC controls industry, which has brought new challenges for both building operators and HVAC controls technicians. Not only are these technicians expected to be technically proficient in network troubleshooting, they must also have the people skills needed to work successfully with new colleagues—the IT staff.

The roles of building operators, HVAC controls technicians, and IT staff can frequently overlap, but their goals differ, creating the potential for tension and barriers to cooperation. The operator and/or technician is responsible for the building operation and occupant comfort, but the IT staff is responsible for information accessibility and security. At times, these objectives can be at odds, and personnel on each side can interfere with the other's work, causing problems or delaying solutions.

However, there are usually ways to reach compromises. An effective collaboration between these groups involves mutual understanding of each other's responsibilities, respect for each other's areas of expertise, and a willingness to cooperate.

Often, building operators or HVAC technicians can solve simple network problems on their own. However, when a complex situation arises, effective collaboration with the IT staff often helps diagnose and resolve the problem quickly.

Building Operators

A *building operator* is a day-to-day building system manager who is responsible for all HVAC, electrical, plumbing, and life-safety systems. The building operator monitors the HVAC control system and makes changes when needed to provide comfort while maintaining energy efficiency. Building operators are typically based on-site and perform tasks such as responding to alarms, modifying time schedules, looking at system performance through trend data, changing setpoints, and overriding equipment controls as needed.

The scope of other building operator duties may vary widely. Most building operators simply manage the HVAC control system that has already been installed by an HVAC controls contractor. However, in some buildings, building operators may also be responsible for maintenance, troubleshooting, additions, and upgrades to the control system. In these cases, building operators may need some of the same training, tools, and resources of an HVAC controls technician.

HVAC Controls Technicians

An *HVAC controls technician* is a technician that installs, configures, maintains, and troubleshoots the equipment and electronics of HVAC system controls. These technicians are expected to know not only HVAC mechanical systems, such as refrigeration systems, boilers, blowers, and electrical systems, but also the complex interoperation of all these systems together and how they affect each other.

Technicians sometimes work on old control systems that use pneumatic or simple electronic signaling. This is challenging in its own way, as the equipment is obsolete and no longer supported by manufacturers. However, technicians now regularly work with new HVAC equipment that includes embedded computers in order to add sophisticated control and data-handling capabilities. **See Figure 1-1.** Also, these systems require supporting network equipment, such as routers and servers. Today's HVAC controls technician must now have a basic knowledge of network architecture, hardware, software, servers, diagnostic and troubleshooting tools, wireless technology, and common networking problems.

Also, network technology often changes, with new products, software updates, and upgrades of existing systems introduced several times a year. In many cases, the product documentation and standard training does not keep pace with these changes. Technicians must be proactive in their own professional development. This may include taking courses in networking or specific equipment, studying technical literature, or consulting online resources.

Computer-Based Controls

Figure 1-1. It is increasingly common for new HVAC equipment to contain computer-based controls and network connections.

IT Technicians

In the case of network-based HVAC control systems, both building operators and controls technicians must work alongside a facility's information technology (IT) staff. An *IT technician* is an expert in the hardware, software, and configuration of the computer networking infrastructure in a facility. This includes the network cabling and connections throughout the facility, the network devices that manage communications, and the settings for all computers and network devices communicating on the network. **See Figure 1-2.** An IT technician may also be referred to as a network technician. IT staff are highly skilled and trained, with many of them holding advanced degrees and industry certifications.

Network Infrastructure

Figure 1-2. The IT staff is responsible for maintaining, upgrading, and troubleshooting all network equipment and infrastructure in a facility. Much of this equipment is housed in a telecom closet.

While policies and specific staff duties may vary among different facilities, an IT technician has the following general responsibilities:

- design and develop the appropriate network and computer system based on the organization's specific needs
- install all network hardware and software and make needed upgrades and repairs
- maintain network and computer system security
- ensure correct system operation and data integrity
- collect data continuously in order to evaluate network performance and develop ways to improve system performance
- troubleshoot and solve problems quickly when they are discovered

A particularly important staff member is the system administrator, also known as a network administrator. Administrators ensure that critical functions, such as email and data storage, are operating correctly. They also verify that all networked computers are operating correctly. Administrators often supervise other IT staff in the day-to-day tasks needed to support the building network.

One of the most important responsibilities of IT staff is network security. Hacking, viruses, malware, and other threats are high-priority security issues. In response, the IT staff may require certain policies or impose limitations, such as mandating complex passwords or not allowing USB flash drives. This may slow down and complicate service and troubleshooting on the HVAC controls.

It is very important that building operators and HVAC technicians understand the roles of the IT staff and respect their capabilities and expertise. The IT staff is often under enormous pressure to solve major computer problems and network outages quickly. At the same time, HVAC technicians have to realize that the goals of the IT staff are not always the same as their own. For example, the IT staff often has network security as a primary goal. The HVAC technician, however, may need to override security policies in order to keep the network HVAC controls up and running.

The IT staff is likely not familiar with HVAC systems. The impact of equipment failure or controller malfunctions may not be apparent to them. This is especially true in troubleshooting situations where the HVAC staff has to respond quickly, but the IT staff has other priorities they consider to be more important.

Effective Collaboration

Typically, the building operator or HVAC technician is expected to conduct some initial network troubleshooting before contacting the IT staff for more advanced problems. Then, the building staff must work with the IT staff to solve network problems affecting the HVAC control system. Despite not being formally trained in IT technology, a building operator or controls technician must have enough knowledge in networking technology and devices to effectively communicate and collaborate with IT staff. It can be frustrating for IT staff to try to solve a problem with a technician who does not understand their terminology and scope of work.

It is critical that the HVAC technician cultivate a professional and positive working relationship with IT staff. With positive communication, the staff can provide critical assistance when troubleshooting and help solve important network problems quickly. Technicians with a high-level of both technical and people skills are in short supply in the HVAC industry and have more opportunities for higher pay and career advancement.

Review Questions

1. What new skills are expected of building operators and HVAC controls technicians working with modern HVAC systems?
2. What are the typical primary duties of a building operator?
3. What is the scope of HVAC system knowledge expected of an HVAC controls technician?
4. What are the special responsibilities of a system administrator?
5. How much network troubleshooting is expected of a building operator or HVAC controls technician?

Networking Overview 2

Modern building control systems often use computer networks to share information between control devices. By far, the most common type of network uses the Ethernet protocol suite, which defines each aspect of how data messages are packaged, signaled, routed, and received. HVAC technicians must be well-versed in Ethernet terminology, procedures, and hardware.

NETWORKING HISTORY

After World War II, the United States government had military facilities and research laboratories scattered across the country, and officials realized that it would be a great advantage for those facilities to share data. Over a period of years, the Advanced Research Projects Agency (ARPA) developed a network of telephone lines that connected a relatively small number of facilities across the United States. Much of the technology now recognized as the Internet was developed by this project.

Gradually, more devices were added to this data network, including devices for private industries, and the commercial Internet began. Since then, the number of users and devices on the Internet, as well as the number of functions, has grown dramatically.

NETWORKING

A *network* is a system of nodes and other devices that are joined using connecting media that provides the means of data communication between the nodes. A *node* is a networked computer (sometimes a very simple one) with a unique network address that sends and/or receives data. Nodes may also be called "hosts." Examples of common nodes are computers, network-enabled printers, controllers, and sensors. Other network devices, such as routers and hubs, direct and distribute the traffic between the nodes. Some of these devices also function as nodes themselves.

Network Architectures

Networks may exist entirely within a building or between devices in different buildings or even in remote locations. Networks may be local-area networks (LANs) or wide-area networks (WANs). **See Figure 2-1.**

A *local-area network (LAN)* is a communication network that spans a relatively small area. A LAN normally encompasses one business or building and does not require the linking of a large number of devices in remote locations. Networked HVAC control systems typically form a LAN.

A *wide-area network (WAN)* is a communication network that spans a relatively large area. A WAN is used to pass information across long distances to multiple geographical locations and potentially to tens of thousands of devices. Large organizations commonly have a WAN to manage communication between multiple LANs.

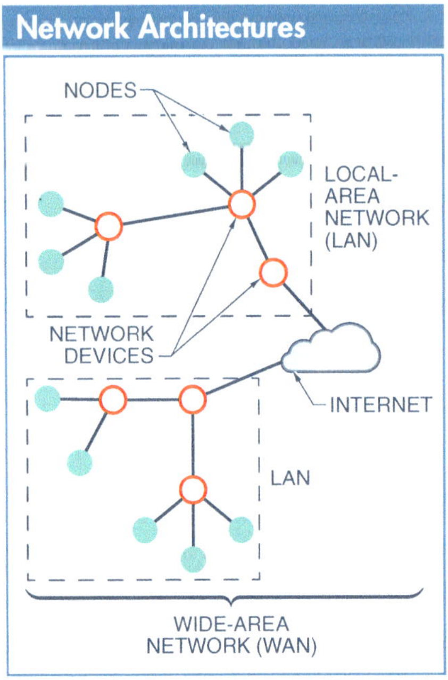

Figure 2-1. Network architectures consist of local-area networks (LANs) that may or may not be connected to other LANs or the Internet to form wide-area networks (WANs).

Network Topologies

A network may be designed in various topologies. A *topology* is a network arrangement that configures the flow of message traffic. **See Figure 2-2.** In a star topology, each node is connected to a central device. In a common-bus topology, each node is connected to a common network wire and not directly to each other. In a daisy chain topology, each node is connected in a line but without the common bus. Rings are similar to common-bus or daisy chain topologies with connected ends. Mesh networks are common for wireless networks in that each node can connect to one or more of its neighbors in any pattern. Many networks feature more than one topology type. For example, a series of local star topologies are commonly connected to a high speed common bus backbone.

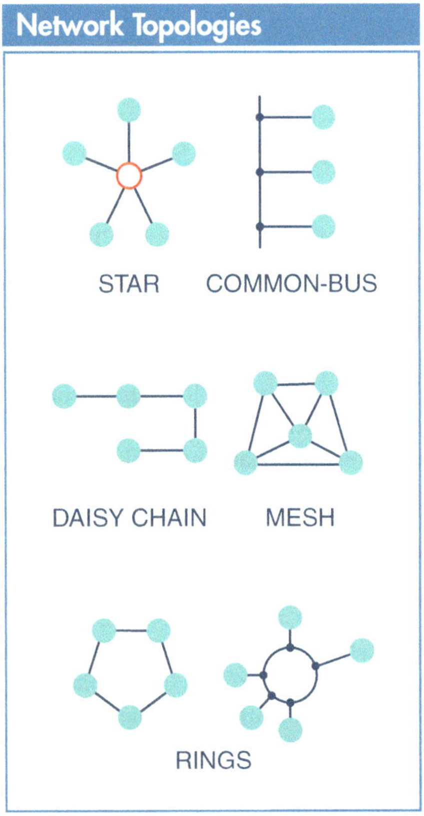

Figure 2-2. A network may be configured in various physical arrangements, known as topologies.

Message Data

All computer data is stored as bits. A *bit* is a unit of information that has only two states, which are commonly labeled as "on" and "off" or "1" and "0." Regardless of the type of data, all information is managed by a computer in the form of bits. When interfacing with humans, however, computers group bits together in ways to represent letters, numbers, colors, sounds, and other data.

Bits are transmitted between nodes with tiny pulses of electricity (or radio signals in wireless communication or light in fiber optic communication) that represent the two states. However, this must be done in an organized manner in order to be useful.

The messages that must be shared between nodes must first be formatted and packaged for transmission. A *packet* is a sequence of bits consisting of a message surrounded by additional information needed to facilitate the transmission and reception of the message. **See Figure 2-3.** This information includes the addresses of the destination and source nodes, the message length, and error detection and correction data. Networks that transmit data arranged in packets are called packet-switched networks.

Networks limit the size of individual packets to help manage traffic. If a message is too long to send as one packet, it is broken up into smaller packets, with information included to help the receiving node reassemble the message.

Bandwidth and Throughput

The individual bits that make up packets are transmitted using changes in the voltage applied to the wiring by the sending node (or changes in radio frequency for wireless networks or changes in light for fiber-optic networks). Each media (wiring) type has a physical limitation to how quickly this signaling can be done reliably. *Bandwidth* is the maximum rate at which bits can be transmitted over a certain media type. However, network communications often do not use all of the available bandwidth. *Throughput* is the actual rate of bit transmission at a specific moment. Bandwidth and throughput are measured in units of bits per second (bps).

Packets

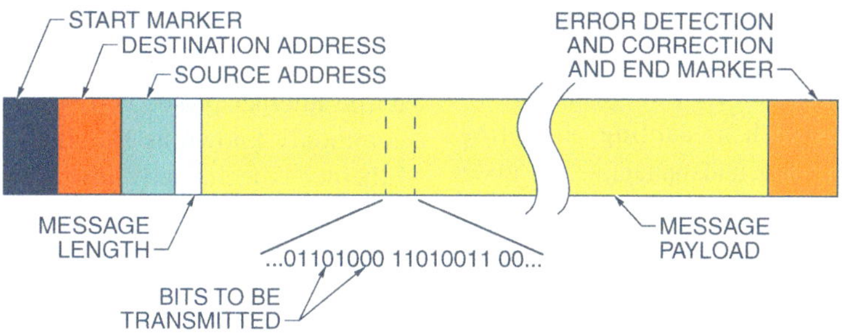

Figure 2-3. Data to be shared between nodes is packaged into units called "packets." Addressing, error correction, and other information is added to the beginning and end of a packet.

Media Access Control

Media access control is the method used by nodes to determine if a network is clear of traffic before transmitting data. The node then transmits its packets of data and waits for a receipt notice from the receiving node. However, if two nodes transmit simultaneously, a collision occurs. A *collision* is a data transmission overlap that corrupts the data. Since the data is corrupted, the receiving node does not acknowledge receipt. Different network types have different methods of media access control and dealing with collisions when they occur.

Addressing

Each node must have an address. An *address* is a node's unique identification code. Address types and formats vary between different network types, but the general concept is universal. Besides nodes, other network devices use addresses to manage traffic and route messages to their destinations. Addressing is critical to network performance.

OSI MODEL

Over the years, many different network types have been developed. Each type of network has its own protocols for functions such as cabling, signaling, packaging data, and managing how nodes take turns in communicating. A *protocol* is a standard or defined procedure that describes a specific communication network function.

The *Open Systems Interconnection (OSI) model* is a way of standardizing the division of a communication system's internal functions by separating it into seven different layers. **See Figure 2-4.** Each layer represents one aspect of the communication process, which is then defined by one or more permitted protocols. These individual protocols may or may not be unique to the overall communication system.

> **ARCNET**
>
> ARCNET is another network architecture that was relatively common in years past but is now considered obsolete. Ethernet is now the standard network type by far. However, HVAC technicians may encounter ARCNET systems that are being serviced or replaced. These systems use similar types of hardware, such as cabling, hubs, and switches, but operate in a very different way, particularly in the way it manages media access. HVAC technicians will require additional training or research in order to work on these systems.

Each data communication system consists of a collection of certain protocols. These collections are often known as a protocol family or protocol suite. The most common data communication system used in building networks is the Internet protocol (IP) suite (also commonly known as Ethernet, though Ethernet is technically just a subset of the IP suite). Many other systems exist for other applications. There are many differences between the various systems. Not all systems utilize every layer, but all conform to at least some aspects of this model. Troubleshooting network hardware and communication problems often involves references to certain layers.

<table>
<tr><th colspan="5">Open Systems Interconnection (OSI) Model Layers</th></tr>
<tr><th>Layer</th><th>Name</th><th>Function</th><th>Related Network Devices</th><th>Internet Protocol Suite Example*</th></tr>
<tr><td>1</td><td>Physical</td><td>Transmits and receives individual bits on the connecting media; involves cabling, voltages, connectors, and pinouts</td><td>Hub</td><td>10/100/1000 BASE-T, Ethernet</td></tr>
<tr><td>2</td><td>Data Link</td><td>Manages physical addressing (MAC addresses) and orderly access to the connecting media</td><td>Switch</td><td>MAC, CSMA/CD</td></tr>
<tr><td>3</td><td>Network</td><td>Manages logical addressing and determines routing between nodes</td><td>Router</td><td>IP, DHCP</td></tr>
<tr><td>4</td><td>Transport</td><td>Provides error detection and correction, segmentation, and reliable end-to-end transmission</td><td rowspan="4">Gateway</td><td>TCP, UDP</td></tr>
<tr><td>5</td><td>Session</td><td>Manages node-to-node dialogs and synchronizes data transfers</td><td rowspan="3">HTTP, FTP, DNS, Telnet, SSH, POP, SMTP</td></tr>
<tr><td>6</td><td>Presentation</td><td>Converts codes, encrypts/decrypts, and reorganizes data; translates between data formats</td></tr>
<tr><td>7</td><td>Application</td><td>Interfaces with user's application program</td></tr>
</table>

** Only selected protocol examples are listed.*

Figure 2-4. The OSI model divides an entire data communication process into seven layers of functions. Networking systems use individual protocols to provide those functions.

ETHERNET NETWORKS

Ethernet is a LAN architecture that can connect up to 1024 nodes and supports minimum bandwidths of 10 Mbps. High-speed networks of 100 Mbps, 1000 Mbps (1 gigabit), and higher are common. Ethernet is used in over 90% of all data communication networks.

A telecom closet (also known as a server room) is the central location for network equipment. Here, devices and cabling are organized into racks and enclosures.

Media Types

The first and most common network component that a technician encounters is the connecting media. *Media* is the cabling that connects the nodes to form the network.

Ethernet can be operated with twisted-pair, coaxial, and/or fiber optic cabling. Specific physical layer standards for Ethernet are identified with a three-part designation. The first part is the communication speed in megabits per second (Mbps). The second part is "BASE," an abbreviation for "baseband." *Baseband* is a data transmission method where data is sent without frequency modulation. This means that there is only one data channel on a cable. The third part is an indicator for the type of cable or other designation.

Twisted-Pair Cables. A *twisted-pair cable* is a cable that consists of two or more pairs of wires, with each pair twisted individually. Twisted-pair cables vary depending on the number of pairs and the type of shielding. **See Figure 2-5.**

Shielding is metal foil or braided wire that surrounds wires in order to prevent electromagnetic interference. Shielding may be around individual pairs and/or around the whole bundle of pairs (just under the protective jacket). For example, shielded twisted-pair (STP) cabling has metal foil wrapped around each pair. Screened twisted-pair (ScTP) cabling has foil or a metal braided screen around the entire bundle. Screened shielded twisted-pair (SSTP) cabling has both types of shielding.

Unshielded twisted-pair (UTP) cabling has no electromagnetic shielding. This has been the most common type of cabling installed in residential and commercial buildings for many years. Common Ethernet UTP cabling standards include 10BASE-T (10 Mbps), 100BASE-TX (100 Mbps), and 1000BASE-T (1000 Mbps or gigabit).

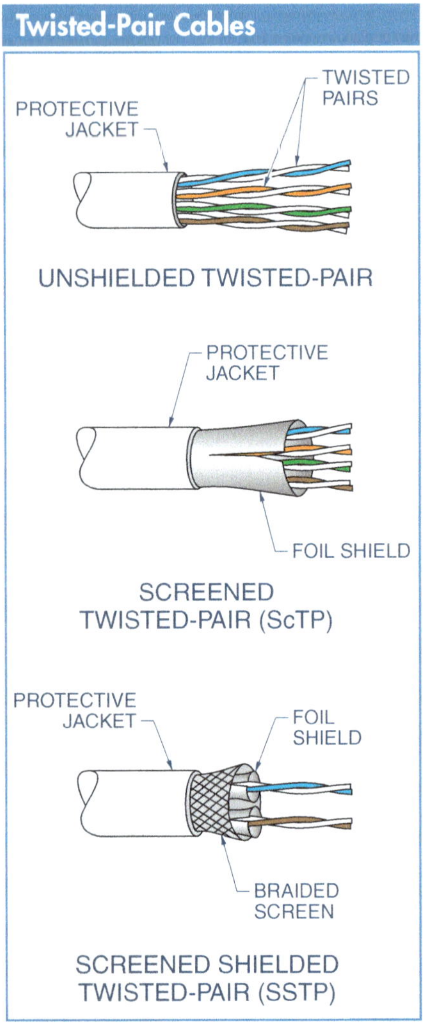

Figure 2-5. Network cabling within a building uses twisted-pair cables. Most installations use unshielded twisted-pair (UTP), though types of shielded cables are becoming more common for their high-bandwidth capabilities.

Twisted-pair cable is classified by a category (CAT) rating that indicates the required signaling performance of the media. See **Figure 2-6.** This rating is printed on the cable jacket or outer insulation. Network wiring specifications list the cable CAT rating required for a specific network. Today, CAT 5, CAT 5e (for "enhanced"), and CAT 6 are the most common cable types. CAT 7 and CAT 8 cables are now being introduced. These higher CAT rating standards require shielded cable in order to provide greater signaling performance. This is making shielded cable gradually more common.

Cables are backward compatible with lower CAT ratings, meaning that higher rated cable can be used with devices that require lower-rated cable as a minimum. However, lower-rated cables may cause communication problems when used with devices that require higher-rated cabling. Therefore, new cabling installations will often use cabling with a higher-than-required CAT rating to prolong their useful life. It is extremely difficult to replace cabling inside of walls and through raceways years later.

Twisted-Pair Cable Category Ratings

<table>
<tr><th>CAT</th><th>Maximum Signal Frequency*</th><th>Maximum Bandwidth†</th><th>Applications</th><th>Note</th></tr>
<tr><td>1</td><td>1</td><td>1</td><td rowspan="2">Voice only</td><td>Obsolete; also known as level 1</td></tr>
<tr><td>2</td><td>4</td><td>4</td><td>Obsolete; also known as level 2</td></tr>
<tr><td>3</td><td>16</td><td>10</td><td rowspan="9">Voice and data</td><td>Common for early data networks; no longer used</td></tr>
<tr><td>4</td><td>20</td><td>16</td><td>Improvement over CAT 3, but quickly surpassed by CAT 5/5e; never commonly used</td></tr>
<tr><td>5</td><td>100</td><td>100</td><td>Superseded by the improved CAT 5e version</td></tr>
<tr><td>5e</td><td>100</td><td>1000</td><td>Very common in recent and new installations due to better performance and similar price compared to CAT 5</td></tr>
<tr><td>6</td><td>250</td><td rowspan="4">10,000</td><td>Becoming the most common cable for new installations</td></tr>
<tr><td>6_A</td><td>500</td><td>Carefully controlled twists in each pair to prevent electromagnetic interference</td></tr>
<tr><td>7</td><td>600</td><td rowspan="2">Must be fully shielded cable</td></tr>
<tr><td>7_A</td><td>1000</td></tr>
<tr><td>8</td><td>2000</td><td>40,000</td><td>Designed for data center installations</td></tr>
</table>

** in MHz*
† in Mbps; also, bandwidth varies depending on signaling standard and segment length.

Figure 2-6. Twisted-pair cables are rated according to their transmission bandwidth, resistance to interference, and other factors. Category (CAT) 5e and 6 are the ratings most commonly installed in buildings.

Normally, long-distance cable runs (permanently installed lengths of cable) are set up by electrical subcontractors. However, technicians may need additional short or medium lengths of cables for various reasons. For example, moving a node or computer may require a temporary connection with a new cable.

Coaxial Cables. *Coaxial cable* is cable that consists of an inner conductor surrounded by an insulating layer, surrounded by a conductive shield, and then a protective covering. **See Figure 2-7.** There are many variants of coaxial cable with different dimensions and shielding types.

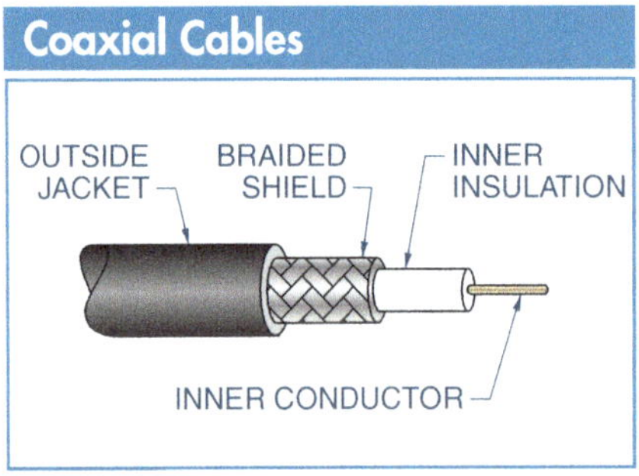

Figure 2-7. Coaxial cable consists of an inner conductor surrounded by insulation and a braided shield.

Ethernet standard 10BASE2 uses a relatively thin coaxial cable designated as RG-58/U, also known as "thinnet." Ethernet standard 10BASE5 uses a thicker RG-8/U cable and is known as "thicknet." RG-8/U is also referred to as "yellow cable" because it is commonly available with a yellow plastic covering. Both cable types were common in early Ethernet networks but became obsolete with the introduction of faster and lower-cost UTP.

Optical Fiber Cables. *Optical fiber cable* is a cable containing glass or plastic fiber used to transmit information using light. **See Figure 2-8.** Optical fiber cables have become popular for WANs and very high-speed Ethernet networks, but they are rarely used in building control systems and must be installed and serviced by professionals qualified in their use. Normally, HVAC technicians do not troubleshoot or install optical fiber cables. If a problem is suspected with optical fiber cables, the facility IT staff should be informed.

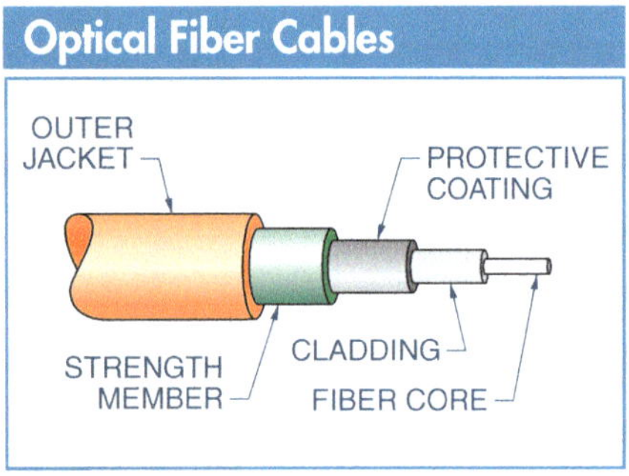

Figure 2-8. Optical fiber cable consists of a fiber core surrounded by various layers to protect and support the core.

Backbone and Horizontal Cabling

Network cabling inside a building is typically classified as backbone cabling or horizontal cabling. **See Figure 2-9.** *Backbone cabling* is network cabling that runs from the entrance facility (where the telecom utility brings in Internet) to and between one or more telecom closets. The backbone system usually incorporates high-speed cabling and equipment for high traffic. *Horizontal cabling* is network cabling that runs from a node or wall jack to the telecom closet.

Backbone and Horizontal Cabling

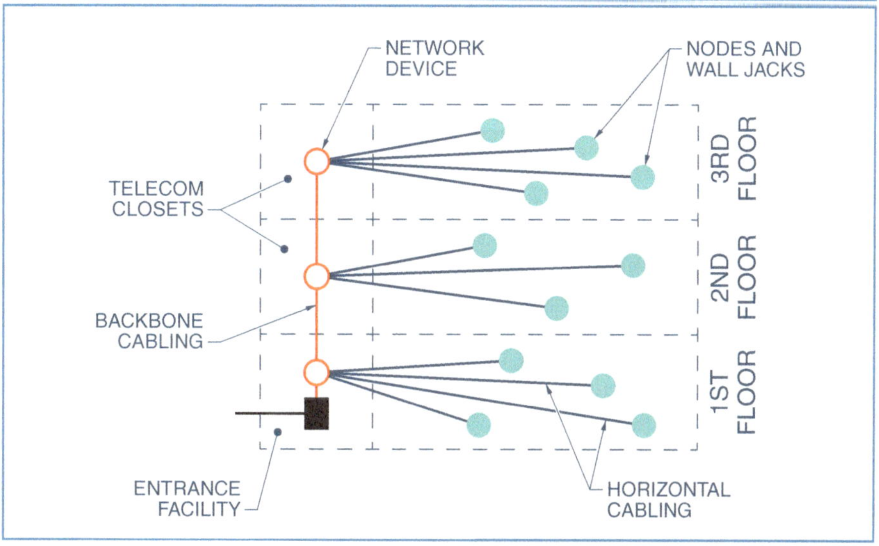

Figure 2-9. Building cabling is typically organized into backbone and horizontal cabling.

A *telecom closet* is a small room dedicated for housing most network devices and being the collection point of network cabling from a building area. These rooms may also be called server rooms, IT closets, or another similar term. The telecom closet should be centrally located for the area it serves, which may be a floor, an entire building, or some other logical zone. Telecom closets usually include one or more racks, which are metal frames designed to hold telecommunication equipment.

Also included in racks are patch panels. A *patch panel* is a rack-mounted faceplate with many network jacks, each of which terminates a cable run from another part of the building. **See Figure 2-10.** Cabling from nodes and wall jacks are all run back to the telecom closet, where they are connected, as needed, to the equipment.

Patch Panels

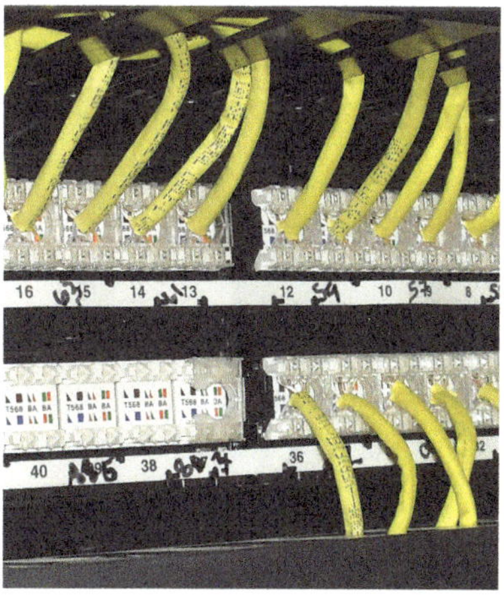

Figure 2-10. Cable runs from wall jacks and certain nodes are terminated at the back of a patch panel.

Patch cables are used to make the final connections. **See Figure 2-11.** A *patch cable* is a relatively short cable used to make connections between two jacks within a telecom closet. Patch cables connect either two nearby network devices or certain wall jacks (via a patch panel) and a nearby network device.

For example, a wall jack in Office 21 is connected via in-wall cabling to a telecom closet down a hall on the second floor of a building. That cable enters the telecom closet and is terminated in a patch panel that also terminates cables from other offices on the second floor. A patch cable then plugs into the jack for Office 21 and the other end plugs into an available jack on a switch in the same rack below it. This cable only needs to be a few feet long. Also, the switch may then be connected with a patch cable to another network device, such as a router.

This connecting scheme has significant advantages. First, there may be more wall jacks installed in a building than are needed at all times. The patch panel is used to connect only the necessary wall jacks to the network devices. Second, if changes are needed to the network, they can be made relatively easily by moving or adding patch cables. If wall jacks were connected directly to network devices instead, upgrading or modifying the network would be far more difficult. However, if not managed carefully, this cabling scheme can result in disorganized telecom

Horizontal Cabling

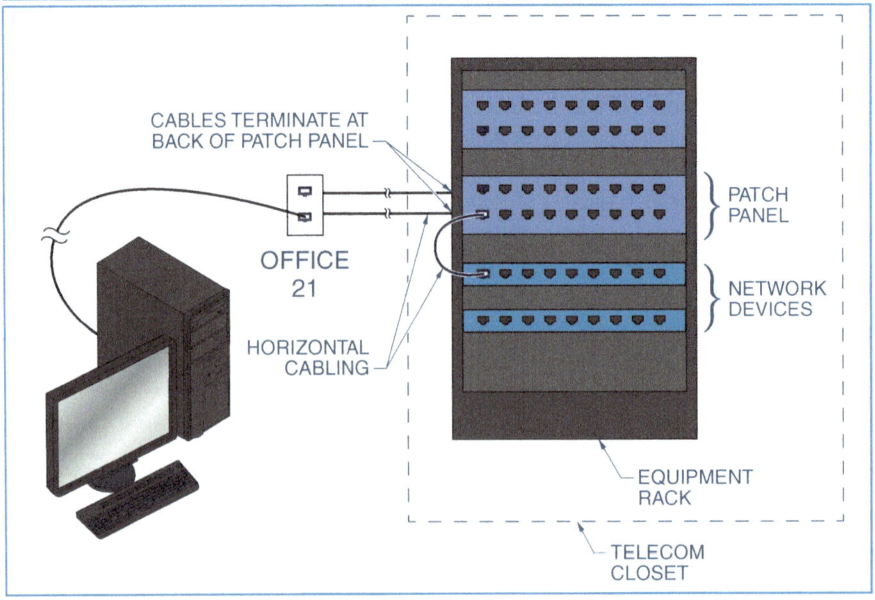

Figure 2-11. Patch panels are used to collect all cable runs from wall jacks and nodes into one organized connection point. Patch cables then make connections between patch panel jacks and other equipment.

closets with tangled and impossible to trace patch cabling. Technicians should use only the cable length needed and, preferably, route the cable along guides on the equipment racks to keep a neat and orderly appearance.

Cable Terminations

Twisted-pair cables are terminated with rectangular plastic connectors designated as RJ-45 connectors. They resemble oversized phone jack connectors. **See Figure 2-12.** An RJ-45 connector contains eight pins instead of the four pins in standard phone connectors. Cables already terminated with RJ-45 connectors can be purchased commercially in various standard lengths. HVAC technicians should have some of these available as they work in the field. Cables can also be easily fabricated in the field using bulk cable spools, modular connectors, and a simple crimping tool.

Patch cables are terminated using the same pin-outs on each connector. Another type of cable is a crossover cable. A *crossover cable* is a cable that is terminated with some pin-outs reversed on each connector. **See Figure 2-13.** The transmission wires from one connector are wired as the receive wires of the other connector. Crossover cables are used for connections directly from node to node. For example, they may be used for temporarily connecting a laptop directly to an HVAC controller to configure or troubleshoot the controller.

Almost all cables are patch cables, and crossover cables are fairly rare. However, the two cables are not interchangeable and may cause a communication failure if misapplied. The types should be labeled for easy identification. A certain cable color, such as blue, is often reserved for crossover cables, but this may not always be the case. Testers can be used to verify the cable type.

RJ-45 Connectors

Figure 2-12. Twisted-pair cables are terminated with RJ-45 connectors, which resemble telephone connectors but are larger and have eight pins.

Connectors can be used to extend the length of twisted-pair cables, but they should be limited to temporary or troubleshooting purposes because each connection is a potential source of signal quality problems.

Terminated Twisted-Pair Cables

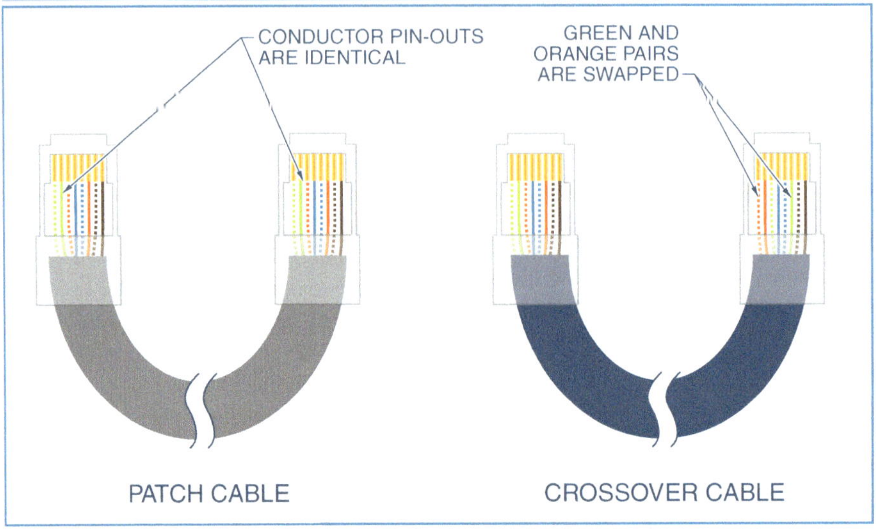

Figure 2-13. Most Ethernet cables are patch cables, though crossover cables are used for certain connections. The only difference is the arrangement of the conductors in the connector.

Network Interface Controller (NIC) Cards

A *network interface controller (NIC) card* is a card (a circuit board with external ports) installed in a node that allows the node to communicate over a network via a particular type of media. **See Figure 2-14.** These are also known as network adapters. For example, a NIC card may provide a computer with a 100BASE-T port. A *port* is a physical connector built into a device to allow the connection of a particular type of cable.

NIC cards are usually included and preinstalled on new computers. On some computers, a NIC is built into the mother board and is not a separate card. Also, some computers have multiple NICs, so technicians must be aware of which port to use.

Network Interface Controller (NIC) Cards

Figure 2-14. A network interface controller (NIC) card allows a node to communicate over a network via a particular type of media.

> The basic concepts and rules of data communication are common to all computer-based networks. This includes office LANs composed of personal computers and HVAC control networks composed of web-based supervisory controllers. Networks are as small as two connected devices or as large as the Internet.

Networked HVAC controllers also utilize NIC cards, though the cards may be proprietary to a manufacturer. Also, the cards may be available separately, so controllers can be set up for different media by selecting a particular NIC card.

Media Access Control (MAC) Addresses

Each node has a media access control (MAC) address. A *media access control (MAC) address* is a unique and permanent address that is hard-coded, or permanently saved, into the device by the manufacturer. MAC addresses are specific to each device and must be unique. MAC addresses are typically formatted as a series of numbers and letters separated by hyphens or colons, such as 00-04-37-E2-DC-25 or 00:04:37:E2:DC:25. The numbers are represented in hexadecimal notation, which includes letters as numerical symbols.

HEXADECIMAL NUMBER SYSTEM

While nearly all raw communication between computer-based devices is in binary notation, these sequences of numbers can be very long and difficult for a person to interpret. Sometimes the information is represented in hexadecimal notation, which shortens the sequences and makes the codes more recognizable.

A hexadecimal number is a number expressed with a base of 16. Since sixteen different symbols are required for this system, the digits 0 to 9 are insufficient. The letters A, B, C, D, E, and F, either uppercase or lowercase, are used for the numbers 10 through 15. Just as in the decimal system, each digit position has a value, which is equal to the base (16) raised to the power of the position, beginning with zero. The values of each position, from right to left, are 1 (160), 16 (161), 256 (162), 4096 (163), etc. To convert the number to decimal notation, each hexadecimal digit is multiplied by its position value.

LETTERS REPRESENT NUMBERS 10 THROUGH 15

Symbols: 0, 1, 2, 3, 4, 5, 6, 7, 8, 9, A, B, C, D, E, AND F

Digit Position:	3	2	1	0
Weighted Values:	16^3	16^2	16^1	16^0
	4096	256	16	1
Example Number:	F	1	A	6

Conversion:

$$
\begin{aligned}
6 \times 1 &= 6 \\
A(10) \times 16 &= 160 \\
1 \times 256 &= 256 \\
F(15) \times 4096 &= \underline{61{,}440} \\
& 61{,}862_{10} = F1A6_{16}
\end{aligned}
$$

BASE DESIGNATION

MAC addresses are usually listed on a label somewhere on the surface of the device. **See Figure 2-15.** These must be recorded and saved in network configuration documentation. When managing many devices, an IT technician may use a MAC address barcode to quickly scan many devices. Database software is used to organize the information. MAC addresses may also be read electronically from most nodes, either over the network or via another communication method (such as a serial port connection).

Media Access Control (MAC) Addresses

Figure 2-15. The MAC address for a device is often shown on a sticker on the surface of the device.

Internet Protocol (IP) Addresses

Internet Protocol (IP) addressing is a method of assigning addresses to nodes on a network. Each Ethernet node must have an IP address. There are two different types of IP addresses available, IPv4 and IPv6. An IPv4 address consists of four sets of numbers, 0 through 255, which are separated by periods, such as 192.168.0.250. While almost all current systems use IPv4, IPv6 will be used more in the future. IPv6 allows a much larger number of possible IP addresses as well as other network performance improvements. An IPv6 address consists of eight sets of numbers, 0 through 65,535, which are separated by colons, such as 2001:db8:0:1234:0:567:8:1. An IPv6 address uses hexadecimal notation, so some numbers are represented by letters.

An IP address may be static or dynamic. A *static (fixed) address* is a permanently assigned address. **See Figure 2-16.** This does not mean that it cannot be changed, such as with a MAC address, but that it remains the same until it is manually changed by a technician. A facility's IT staff can determine a method for assigning IP addresses, such as selecting particular ranges for certain node types or locations, and use software utilities to define an address for each node.

A *dynamic address* is an address that can be automatically changed due to network changes. For example, if a node is power-cycled (turned off and then back on), it may be assigned a new IP address. Or a node that is inactive for a long period may have its dynamic IP address expire and must request a new one. Dynamic addressing uses the Dynamic Host Configuration Protocol (DHCP). A *Dynamic Host Configuration Protocol (DHCP) server* is a common router feature that automatically assigns IP addresses to nodes as they are connected to the router and then keeps track of the addresses.

IP Addresses

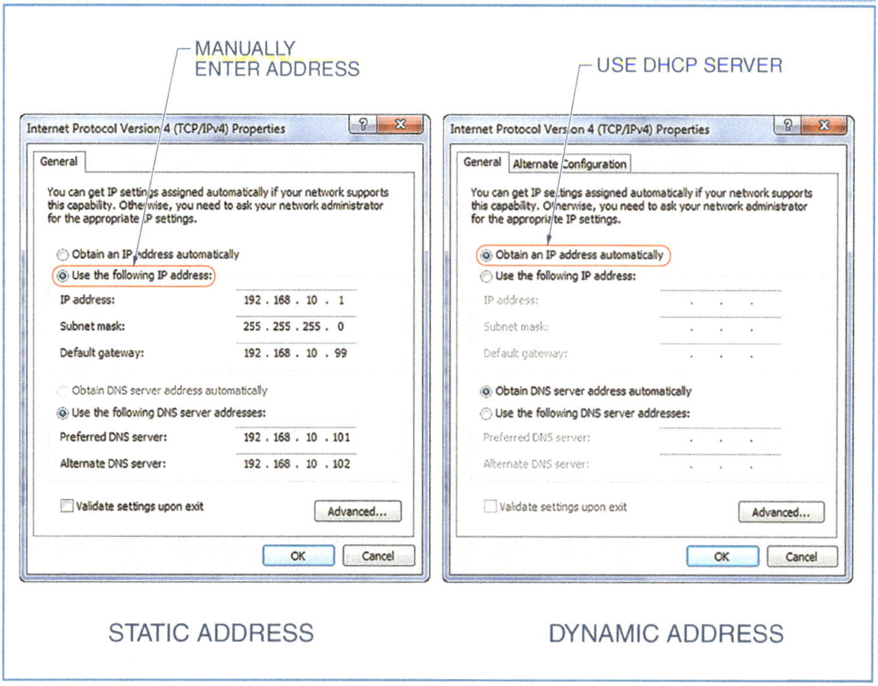

Figure 2-16. A node is either configured with a static (fixed) IP address or uses a DHCP server to automatically obtain an available IP address. Subnet mask, router, and DNS server addresses may be configured similarly.

Subnet Masking. An IP address consists of a network address and a host (node) address. This system is used by routers to easily determine whether messages should remain within the subnetwork or be sent to the main network. For example, for the IP address 192.168.123.250, the network address may be 192.168.123.0, and the host address may be 0.0.0.250. In this case, the IP address is divided at the last period to form the two constituent addresses. However, the dividing point between the two can vary, depending on the network design.

Subnet masking is an addressing scheme that defines the division between a node's network address and host address. The mask has the same format as an IP address and, when displayed in binary form, forms a continuous series of ones followed by a continuous series of zeros. The point at which the ones change to zeroes indicates the division between the network and host addresses. **See Figure 2-17.** Common subnet masks are 255.255.255.0 and 255.255.255.192. Improper configuration of a node's subnet mask can cause network communications involving the node to fail.

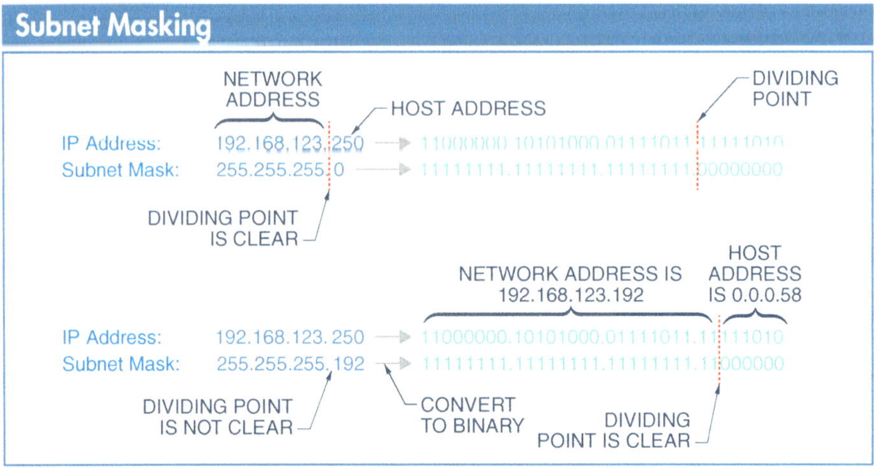

Figure 2-17. Subnet masking is an addressing technique that defines the division between the network address and host address of a node.

BINARY NUMBERING SYSTEM

A binary number is a number expressed with a base of 2. The only two symbols used in this system are 0 and 1. A bit is a binary digit, consisting of a 0 or a 1. For binary notation, the values of each position, from right to left, are 1 (2^0), 2 (2^1), 4 (2^2), 8 (2^3), 16 (2^4), 32 (2^5), 64 (2^6), etc. A 16-bit binary number can represent a maximum decimal number of 65,535.

Communications using binary numbers typically organize the information in groups of 4, 8, or 16 bits. A small number may not require all of the bits in a group to represent its value, but since a certain number of bits may be expected by the message receiver, leading zeroes are sometimes added. For example, the decimal number 18 can be expressed in binary as 10010, but if 8 bits were reserved to represent this number, the transmitted signal would be 00010010.

Symbols: 0 and 1

Digit Position:	7	6	5	4	3	2	1	0
Weighted Values:	2^7	2^6	2^5	2^4	2^3	2^2	2^1	2^0
	128	64	32	16	8	4	2	1
Example Number:	1	0	1	1	0	1	1	0

Conversion:

$$
\begin{aligned}
0 \times 1 &= 0\\
1 \times 2 &= 2\\
1 \times 4 &= 4\\
0 \times 8 &= 0\\
1 \times 16 &= 16\\
1 \times 32 &= 32\\
0 \times 64 &= 0\\
1 \times 128 &= 128\\
\hline
& 182_{10} = 10110110_2
\end{aligned}
$$

DNS Servers. Network devices can be identified either by computer name or IP address. If DHCP is used, the IP address is not fixed, so the device may need to be identified by its computer name, or hostname, for addressing. A *hostname* is a unique node name that is human-readable. For example, a local network hostname might be "AHU2-controller." An example of an Internet hostname is "www.example.com." A *Domain Name System (DNS) server* is a computer server or router function that translates hostnames into IP addresses so that messages can be routed properly.

If a DNS server is not functioning properly, the network may not be able to use hostnames. This problem may be indicated with a "Cannot resolve hostname" error from the browser. Setting up and troubleshooting a DNS server is the responsibility of the IT staff and not the HVAC technician.

Ports. Sometimes a network message must be directed to a certain application program on a node, so a node-only address is not sufficient. In this case, a port number is added to the node's IP address to identify the application destination. A *port* is a virtual communication connection used by nodes to exchange data directly with certain application programs on other nodes. (This is different from a physical port used to connect a cable to a device.) Common network ports include 25 for Simple Mail Transfer Protocol (SMTP), 80 for Hypertext Transfer Protocol (HTTP) used with the Internet, and 443 for HTTP Secure. Port numbers are appended to IP addresses after a colon. For example, a port-specific IP address may be 192.168.123.250:25.

An HVAC technician may need to know which ports are needed for the control system and inform the IT staff so the ports are open and not blocked by a firewall. Otherwise, devices may not communicate properly. The controller manufacturer software or documentation can be consulted for a list of ports.

Network Devices

For two nodes to communicate with only each other, only a single connecting cable is necessary. However, a network device is needed when more than two nodes need to communicate. A *network device* is a type of switchboard that allows three or more nodes to communicate with each other. Each node is connected to a network device, which sends the messages from one node to another node. Network devices also help manage network traffic and prevent network problems. Common network devices include hubs, switches, and routers.

Hubs. The simplest and least expensive network device is the hub. A *hub* is a multiple-port network device that sends all messages received from any port back out to every port. **See Figure 2-18.** It consists of four or more ports, a power supply, and lights that indicate network activity. A hub passes an incoming message to all of the other ports indiscriminately. Hubs are plug-and-play components, meaning that no configuration is normally needed. With today's networks, hubs are not used as much as switches.

Switches. A switch is more sophisticated than a hub. A *switch* is a multiple-port network device that uses MAC addresses to forward network messages to only the port of the intended recipient. **See Figure 2-19.**

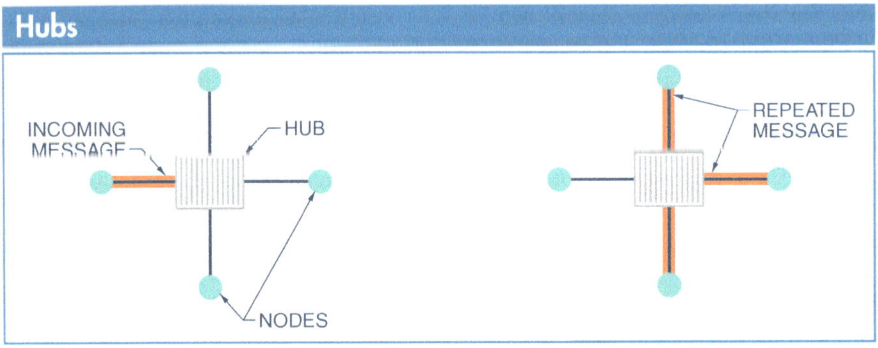

Figure 2-18. A hub connects multiple nodes and repeats all received messages onto its other ports regardless of the destination addresses.

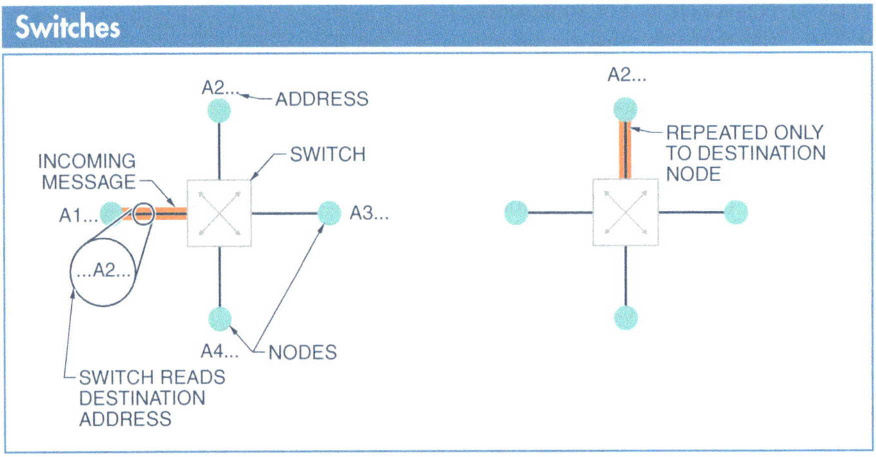

Figure 2-19. A switch connects multiple nodes and selectively forwards messages to specific destination nodes, based on their MAC addresses.

When configured, the switch learns the MAC address of the node connected to each port. It can also inspect each message received for the embedded MAC address of the intended recipient so that it can send the message to the correct node. This greatly reduces network traffic in a manner that a hub cannot do.

Switches also have the ability to automatically change the data transmission speeds between different networks or parts of a network. If these speeds cannot be changed, the entire network transmits at the slowest speed. By reducing traffic and managing speeds, switches can greatly improve network performance over using hubs alone.

Many networked HVAC controller enclosures include a simple four-port switch. **See Figure 2-20.** This allows access to the network by a field technician using a laptop and a patch cable. However, it is advisable for a technician to have a small switch available in case such access is not already available.

Switches in Controller Enclosures

Figure 2-20. Technicians usually use small switches to temporarily connect laptops to web-based controllers.

Switches are available in configurations from four ports to as many as 24 ports. When a switch is purchased, some ports should be left available for future use.

Routers. A router is more sophisticated than either a hub or a switch. A *router* is a network device that manages communication between different sections of a network based on IP address. **See Figure 2-21.** Routers receive messages, read the IP addresses of the destinations, and send them to the appropriate ports. Routers are similar to switches except that they operate at a higher layer with more sophisticated capabilities. For example, if there is more than one route to reach a destination, routers choose the best available path. Also, routers typically perform many other related functions, such as providing firewalls or functioning as dynamic addressing servers.

Routers and Subnetworks

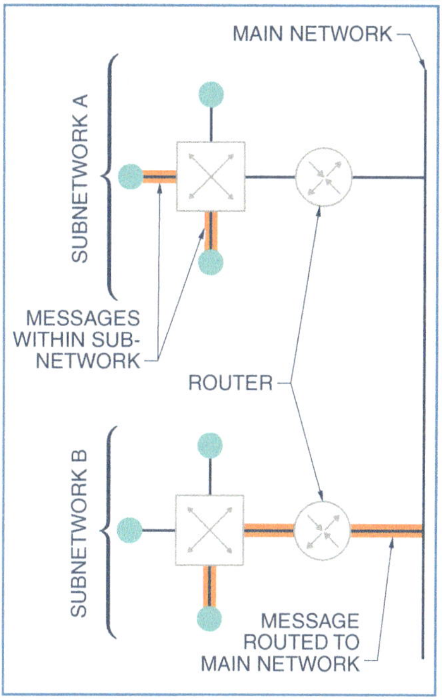

Figure 2-21. A router's primary function is to forward messages selectively based on IP addresses, which allows the router to manage traffic on subnetworks.

Consumer-grade routers typically combine the functions of a router and a switch into a single device, though they are still labeled as "routers." For these small networks, a few ports are adequate. However, large networks need more features and many more ports, requiring separate routers and switches. These switches may have dozens of ports.

A router is typically used to create logical subnetworks on a large system. A *subnetwork* is a subdivision of a larger network that is made up of nodes that communicate frequently. Networks are divided into subnetworks by function and/or location. For example, a logical subnetwork may consist of all of the computers on a certain floor of a building or all of a building's lighting control nodes. A router connects a subnetwork to the main network and allows the transmission of messages to other subnetworks. However, it keeps traffic exclusively intended for its subnetwork within that network, which minimizes traffic on the main network.

Sometimes the term "gateway" is used for a router, but this is not accurate. A gateway is actually a different and more sophisticated network device that has the additional ability to translate between different communication protocols. Care should be taken to use these terms correctly.

Routers have their own IP and MAC addresses. The IT staff uses software to configure routers. Likewise, each node is configured to communicate with a specific router. Therefore, if a router is replaced or reconfigured, this can easily affect or disrupt communication between certain nodes.

A router often includes firewall functions. A *firewall* is a routing function that prevents unauthorized access or the transmission of certain message types between the network and the Internet. This is a major security function that keeps outside users or software from accessing or modifying facility data or controls. It also prevents on-site users from inadvertently downloading malicious software. Due to the significant risk of computer viruses from the Internet, firewalls are essentially required for all networks. Firewall settings for a router may be accessed by using a web browser and are generally password protected. Firewalls are configured to refuse to forward messages from or to certain addresses or messages that contain certain data.

Review Questions

1. What is the difference between a LAN and a WAN?
2. How are messages prepared for transmission on a network?
3. Why is media access control important?
4. What is the most common type of network cabling?
5. What are the advantages of using patch panels to connect nodes and network devices?
6. What is the difference between a static IP address and a dynamic IP address?
7. What is subnet masking?
8. Why are switches generally better and faster than hubs at operating a network?
9. What information does a router use to manage messages?
10. What are subnetworks and how are they defined?

3 HVAC Control System Networks

Web-based supervisory controllers are the apex of an HVAC control system hierarchy. These supervisory controllers also communicate with and monitor various field controllers via a communication bus. They are also connected via Ethernet, either separate from or integrated with the rest of the building network infrastructure. This allows sophisticated access and control features that are not otherwise possible.

HVAC CONTROL NETWORKS

There are two broad types of network architectures for HVAC control systems. **See Figure 3-1.** Some are designed as a separate network, independent of and operating in parallel to a facility's main computer network. Others integrate the HVAC control devices with the facility's computers as part of the main network.

There are several advantages to a separate HVAC control system network. They are typically easier to troubleshoot because they are limited to a relatively small number of nodes. They have less network traffic to manage, which usually improves the network's reliability. They also eliminate any possible security issues with the main facility network. Security is a major concern for many facilities, particularly government and laboratory facilities. However, installation and maintenance costs for separate networks are usually higher due to the duplication of cabling infrastructure, network components, and software.

Alternatively, integrating the HVAC controllers into a main facility network reduces installation costs due to the consolidation of some network devices and fewer cabling runs. However, this design may also complicate troubleshooting because of the higher number of total devices involved on the network.

In either type of network architecture, the HVAC control system has multiple networked controllers. They communicate with each other as well as client computers, servers, tablets, and smartphones using Ethernet.

> Field devices (inputs and outputs) are connected to field controllers, which may allow various types of data communication, including simple analog or digital signals, or communication protocol signals.

Field Devices

A *field device* is an input or output device. HVAC system inputs include temperatures, pressures, damper positions, and ON/OFF status. These are measured by sensors or switches that collect information about actual system conditions. Outputs to the HVAC system include fan speed, damper position, valve position, and ON/OFF switching. Corresponding field devices include variable-speed drives, actuators, and relays. Field devices are connected to a field controller.

HVAC Control Network Designs

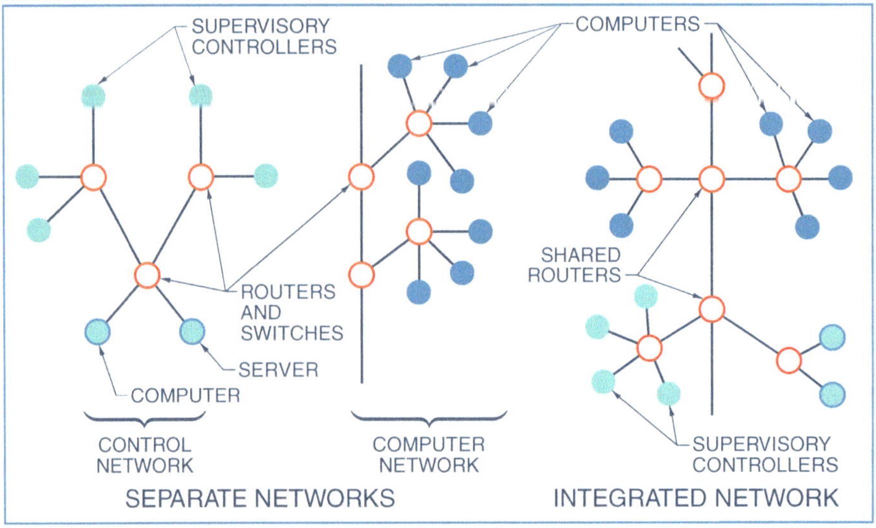

Figure 3-1. HVAC control networks may be either completely separate from or integrated with a facility's main computer network.

Field Controllers

There is typically a dedicated field controller for every large piece of equipment, such as an air-handling unit (AHU), chiller, cooling tower, VAV terminal box, or boiler. A *field controller* is an electronic device that makes control decisions for a specific piece of HVAC equipment based on input information and sends signals as needed to actuators in order to maintain configured setpoints.

These field controllers are connected to field devices: the sensors that collect control information and the actuators that affect the system operation. **See Figure 3-2.** The field controllers also communicate with supervisory controllers via one or more communication buses, such as LonTalk®, BacNet® MS/TP, MODBUS®, and Metasys® N2. These communication buses may be either wired or wireless.

Supervisory Controllers

In a networked HVAC control system, supervisory controllers are connected to the Ethernet network. A *supervisory controller* is an electronic device that contains software-based control functions, coordinates information between field controllers, and may contain high-level control logic. The software-based control functions typically include scheduling, alarming, and data-trending capabilities.

Supervisory controllers are also commonly called front-end devices. Examples of common series of supervisory controllers in HVAC controls include the Tridium® JACE®, Trane® Tracer® SC, and Johnson Controls Metasys Network Automation Engine (NAE).

Each networked supervisory controller is assigned an IP address and subnet mask. Because the supervisory controller is connected to an Ethernet network, it can provide features such as web-based access or emailing. For example, if a temperature exceeds a certain value and triggers an alarm, the supervisory controller can email the building operator or HVAC controls technician. That person may then be able to access the controller remotely to further investigate the problem.

A supervisory controller is typically connected to multiple field controllers. The supervisory controller can then share information between the field controllers. For example, if an indoor temperature sensor indicates that heating is needed, the AHU field controller sets up the heating coil and duct dampers for heating and the boiler field controller receives the signal to start a firing cycle.

Control System Networks

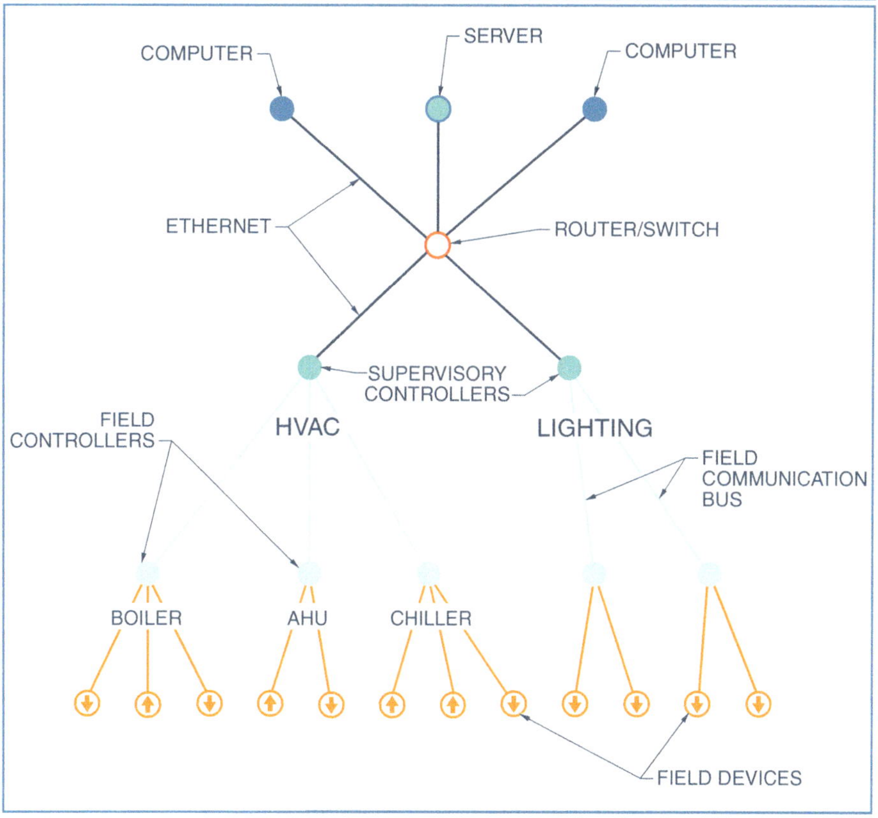

Figure 3-2. Field devices provide input information to and output actions from field controllers, each of which operates a piece of building equipment. Field controllers also communicate with supervisory controllers. Supervisory controllers provide high-level control features and an interface with computers on an Ethernet network.

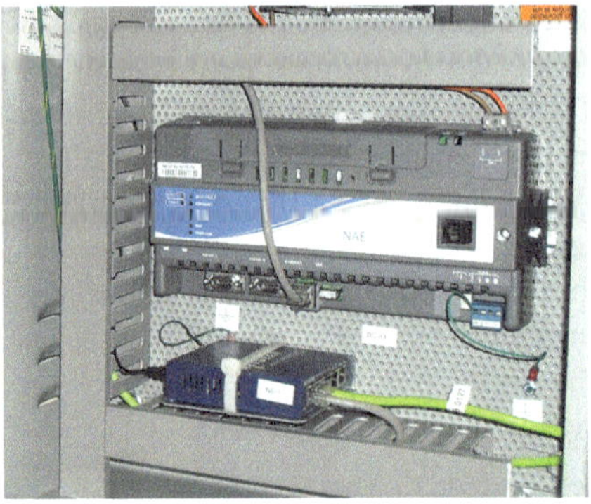

Web-based supervisory controllers are connected to both field controllers and an Ethernet network and are commonly configured via web browser over the Ethernet network.

There may also be multiple supervisory controllers on the Ethernet network. Their responsibilities may be divided by building zone or other logical dividing characteristic. A facility with extensive building automation features may also have additional supervisory controllers for other building systems, such as lighting and access control (door security). If the controllers are compatible, they can share information between them. For example, if a person swipes their identification at a door to unlock it, their office lighting can be turned on and the temperature setpoint changed to an occupied setting.

Web-Based Access

The networked supervisory controller must be configured with both network and HVAC system operating settings. Technicians and facility operators also need to be able to view the HVAC system status and other information. However, it is not cost efficient to build keyboard and display interfaces into each controller. Alternatively, controller manufacturers can develop proprietary software for communicating with a controller from a networked computer, but then there are potential problems of cost, software distribution, and operating system (OS) compatibility. Instead, most network supervisory controllers include a web server function and the programming necessary to deliver an interactive website-like interface to any web browser. These controllers are known as "web-based supervisory controllers." This strategy provides a platform for a rich and flexible interface delivered in a very efficient way.

Similar to other websites, the controller interface is loaded by using the IP address or hostname in the address line of a web browser. For example, an address might be "http://192.168.123.250/". The appearance and functions of the interface depend on the controller manufacturer's design. Typically, though, graphics are used to easily convey status information quickly. **See Figure 3-3.** Colors may be used to highlight normal or out-of-range sensor data. Trends over time may be shown in graph or chart form. Also, animations of moving parts, such as dampers or fans, are commonly included.

Access to web-based supervisory controllers may be limited to only computers on the on-site LAN or available to any computer on the Internet. For security, password-protection and encryption features are typically needed, in addition to special router and firewall configurations for off-site access. A technician or building operator can then use their home PC to access the building network while away from the office. Mobile devices such as smartphones and tablets can also be used to access web-based controllers.

Web-Based Supervisory Controller Interfaces

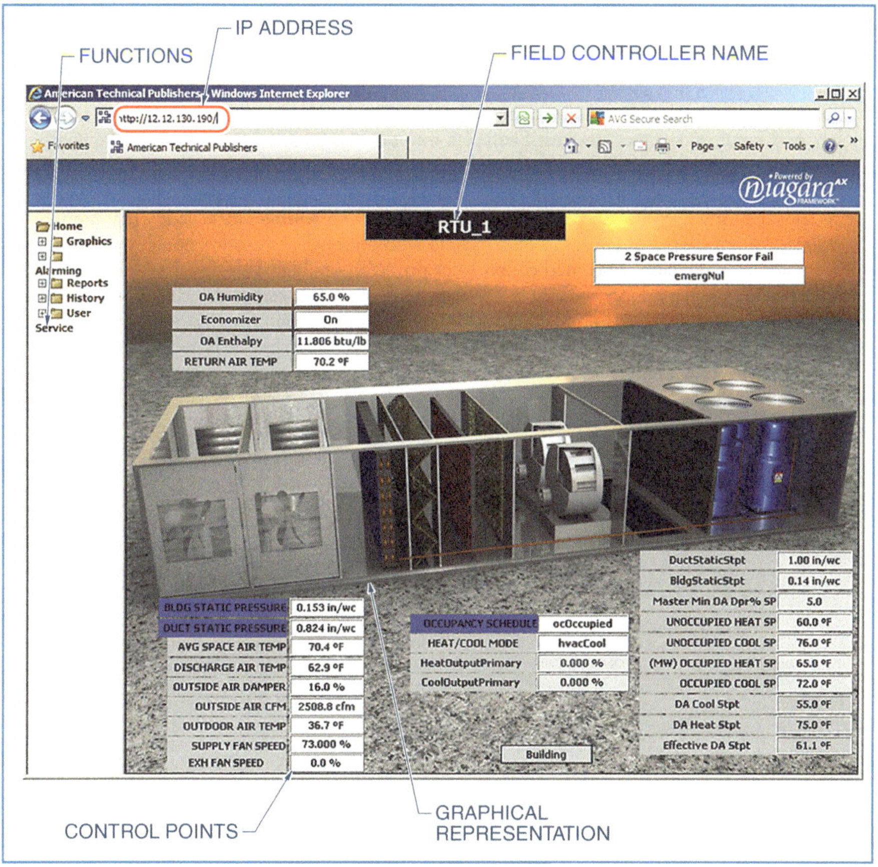

Figure 3-3. A web-based supervisory controller's interface is accessed with a browser application in the form of a web page served from the controller.

WORKING WITH WEB-BASED SUPERVISORY CONTROLLERS

An HVAC controls technician, and perhaps even a building operator, will encounter web-based supervisory controllers when working on modern networked control systems. This is where they must combine their HVAC and IT knowledge and skills.

Controller Installation

One of the common tasks performed by an HVAC controls technician is to install a web-based supervisory controller. This may either be a new installation or an upgrade from an older system. The tools needed include basic hand tools, a laptop, a portable Ethernet switch, patch cables, and any necessary software.

INTERNET OF THINGS (IoT)

Most building controls systems connect to the network via a few devices, such as supervisory controllers, client and server computers, or smartphones and tablets. However, direct network connectivity is now being built into many other HVAC-related devices, including field controllers and input and output devices. Examples include web-enabled environmental (temperature, pressure, and humidity) sensors, relays, and actuators.

This trend is called the "Internet of Things (IoT)" and is changing electronic device ecosystems in many different areas. Smart homes include web-enabled locks, cameras, thermostats, lights, and appliances. Hospitals include web-enabled devices for health monitoring, emergency notification, and medication tracking. Web-enabled vehicles use web-enabled toll collection and parking spaces. In commercial buildings, besides the HVAC system, the IoT may also include web-enabled power meters, lighting controls, and security systems. Through the Internet, all of these web-enabled devices can then interface with other devices, remote users, and even service technicians.

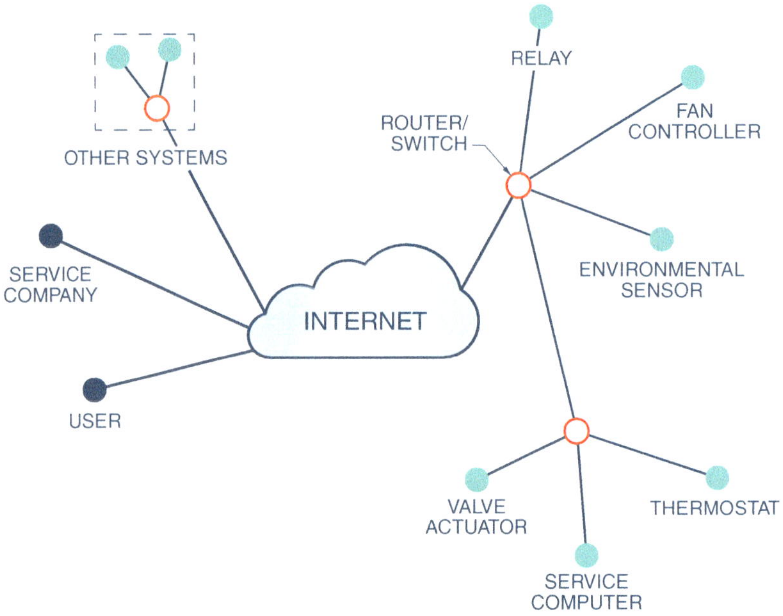

IoT has been made possible by continuing advancements in many related technologies, including miniaturization of electronics, increased router and switch capabilities, improved Wi-Fi coverage, and expanded IP addressing. Version 6 of the IP addressing standard (IPv6) uses 128 bits, which allows for 1028 times more unique addresses than the current (v4) standard.

Wide deployment of IoT leads to "smart buildings," where many different functions can be accessed and controlled via the web. This will increase the need for technicians to both understand and be able to troubleshoot building Ethernet network problems.

A controller is installed using the following procedure:

1. Locate an available enclosure or install a new one. **See Figure 3-4.** It should be in an accessible location and satisfy all device requirements, such as temperature and humidity conditions.
2. Wire the power supply, which is typically a 120 VAC to 24 VAC transformer.
3. Install the web-based supervisory controller in the panel.
4. Obtain an IP address from the IT staff if needed. Otherwise, use DHCP.
5. Attach the Ethernet cable to the controller. The other end should already be connected to the network by the IT staff.
6. Power up the controller.
7. Use a laptop and web browser to access and configure the controller, build the database, add graphics, etc.
8. Use a label-maker to document the IP address, hostname, and subnet mask on the controller.
9. Connect the communication bus from the field controllers to the web-based supervisory controller.
10. Automatically detect and configure the field controllers, as well as their connected inputs, outputs, and parameters.
11. Commission the system by confirming the proper operation of all field controllers.
12. Configure the customer's computers for system and/or individual controller access.
13. Load and/or configure any additional software for the customer's devices, such as smartphones or email alarms.
14. Save copies of the system database on the customer's network and also off-site as backups.
15. Complete the software and device license registration processes with the manufacturers.

Controller Enclosures

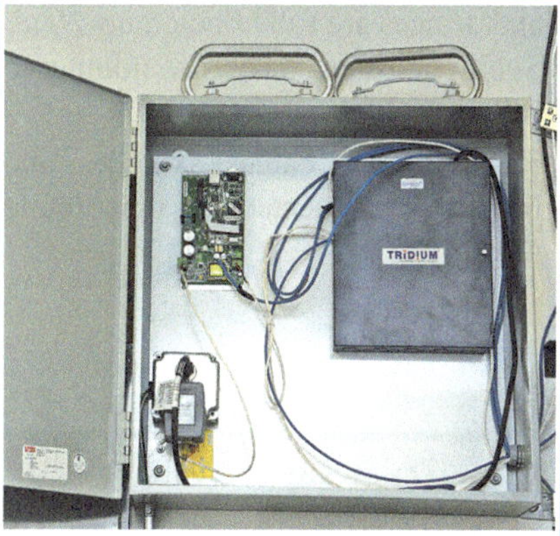

Figure 3-4. Controllers are typically mounted in enclosures that protect the devices but also allow easy access when needed.

Controller Software Updates

Controller manufacturers periodically update the software for their web-based devices, typically about once or twice a year. They add features and capabilities as well as fix problems. A system can be patched or upgraded as needed using a laptop loaded with the manufacturer's software. The manufacturer will provide the software and procedures for the specific controller. In some cases, the entire operating system of the device has to be replaced. This is known as "reimaging" or "flashing" the device.

Basic Controller Troubleshooting

Complex network communication problems should be addressed by the IT staff. However, there are many controller communication problems caused by relatively simple issues that HVAC control technicians can solve. Several simple checks that can be used when troubleshooting common problems include the following:

- Ensure that all power is connected and devices are fully booted up. *Note:* Some controllers take a few minutes to fully initialize.
- Ensure that all cables, switches, and routers are connected and functioning properly.
- Verify that the correct type and version of web browser is installed.
- Disable browser popup blockers.
- Verify that the correct versions of Java®, Flash®, or other necessary supplementary software are installed.
- Use the ping command to verify communication to the IP address.

> When a new HVAC control system is being installed, it may take a few adjustments over a span of time to optimize the field controller settings and the web-based supervisory controller configurations.

Review Questions

1. What are the two broad types of HVAC network architectures?
2. What are field devices connected to?
3. What are field controllers?
4. How do supervisory controllers communicate?
5. What does it mean that a supervisory controller is "web-based"?
6. What are the common tools needed for installing a web-based supervisory controller?
7. What are some software checks that are part of controller troubleshooting?

Client Computers 4

Many different software programs may be involved in setting up and configuring a networked HVAC control system. By far, more problems are caused by the software used with a control system than the hardware. HVAC technicians must be knowledgeable and effective in troubleshooting these components while working with the facility IT staff.

HVAC CONTROL SYSTEM CLIENT COMPUTERS

An HVAC technician typically works on-site on a customer-owned computer that may be dedicated for HVAC control system access. This is known as the client computer. A *client computer* is a networked computer that requests information from a server or other nodes. In this case, the client computer receives webpages from the system server or directly from the web-based supervisory controllers. The computer may be either a desktop or laptop, with laptops favored due to their portability. Technicians may also set up and configure additional computers for system access, as requested by the customer.

Computer Configurations

There are three types of computer configurations for accessing HVAC control systems. The simplest is a browser-only configuration. The user needs only a network connection and a web browser to view HVAC control system information, to issue commands, and to view trends. This requires only a default system setup and a compatible browser, so it usually takes only a few minutes to configure.

Another common configuration requires software tools in addition to a web browser. In this case, a number of controller configuration tools, diagnostic tools, and utilities are installed. The computer requirements are much greater, and there are more opportunities for configuration issues. Also, it may take several hours to load the prerequisite and vendor software. Finally, the third option involves a dedicated server, which will be discussed in a later chapter.

IT Policies

The facility IT staff likely has network hardware and software policies that determine how the network is set up and operates. There may be conflicts between these policies and the requirements of the networked HVAC control system. The IT staff may not be aware of these requirements and potential problems, so it is strongly recommended that technicians discuss the specifics with the staff and provide system documentation, preferably early enough in the project to avoid difficulties. Examples of common policy areas include the following:

- network passwords, for security reasons
- accessing the Internet, for security reasons

- accessing the HVAC control system from outside the network, for security reasons
- installation of software patches and upgrades, for security and software stability reasons
- use of portable storage devices, such as flash drives, for security reasons
- launching specific programs, for security reasons
- installing new software, for security reasons
- changing settings in the operating system, for security and software stability reasons
- documentation of software licenses, for legal and developer support reasons

COMPUTER SOFTWARE

The technician is responsible for any computer problems caused by the software. The technician must consider the intended use of the computer and what software must be installed in order to avoid problems with misconfigurations, version incompatibilities, and software conflicts.

Operating Systems

The single most important software consideration for a computer on an HVAC control system is the operating system (OS). Most HVAC control system clients utilize a version of Microsoft® Windows®. It is absolutely vital that the version of Windows used is compatible with any vendor software needed. Common versions of Windows include:

- obsolete versions, such as Windows XP and earlier
- Windows 7 and 8
- modern versions, such as Windows 10

Each of these OSs may have an additional edition designation, which may be critical for HVAC control system support. **See Figure 4-1.** For example, Windows 10 is available in several editions, including Home, Pro, and Enterprise. *Note:* Different operating systems are available for computers used as servers, which will be discussed in a later chapter.

Service Packs. After the operating system, the next most important consideration is the service pack loaded on the computer. A *service pack (SP)* is a package of additional operating system components and updates added to improve system operation. It is very important to note the service pack version for a computer, as it may affect compatibility with vendor software. Service packs for Windows may be downloaded from Microsoft's website. Technicians should always check HVAC control system software requirements to be sure there are no compatibility issues.

Patches. A *patch* is a software update, typically for operating systems, installed to fix bugs and/or improve system security. OS patches are typically downloaded from the Internet, often automatically, and installed after other OS software. OS patch versions are specific to both the operating system and the service pack installed.

It is possible that an HVAC control system vendor will require certain patches to be installed. This list is often found in the installation instructions. This can be a problem for computers that are never connected to the Internet due to security concerns. In some cases, the computer can be temporarily connected to the Internet in order to download the software patches.

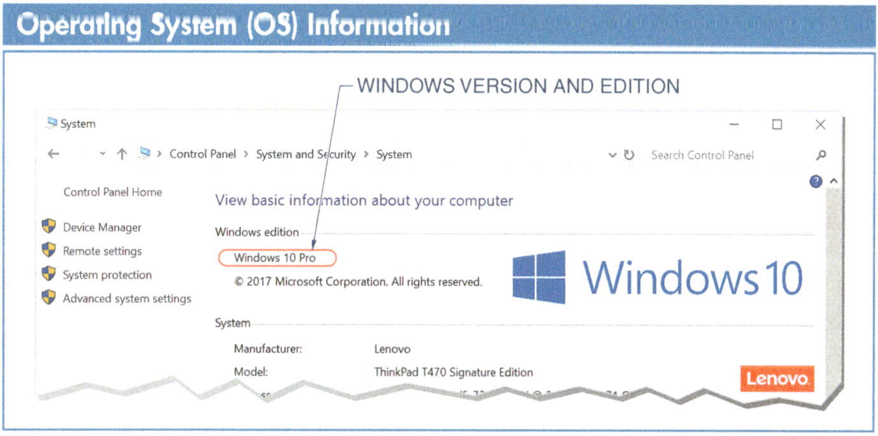

Figure 4-1. The Windows Control Panel can be used to view a computer's OS information, including any service packs that may be installed.

Java

An additional consideration is the use of Java, which is a software environment that allows developers to create application programs that run on a variety of operating systems. Java is used in two ways: to support stand-alone applications and to execute mini-applications, called applets, within a web browser. HVAC controller manufacturers may create custom application programs and interfaces for displaying control system information on computers. By utilizing the Java environment, they can create a single application that is compatible with multiple operating systems and/or browsers. However, not all browsers support Java, so Java compatibility may affect browser choice.

The Java environment and/or browser plug-ins must be installed before a Java application can be used on a computer. Many Java applications require a certain version of the Java environment. **See Figure 4-2.** Also, the required version may vary between different application versions and different manufacturers. For instance, version 5.0 of a controller application may require Java version 8, while version 5.2 of the controller software may require Java 9. For this reason, HVAC technicians should keep all of the appropriate versions of Java on a flash drive and keep the flash drive with them when performing service. Also, it is recommended that technicians maintain a list of controller software versions and their required Java versions for quick reference.

Web Browsers

One of the most important pieces of software on an HVAC control system computer is the web browser. On most control systems, the browser allows technicians and customers to view and interact with the configurations and data. Examples of common browsers include Microsoft Edge®, Microsoft Internet Explorer®, Mozilla® Firefox®, Google® Chrome®, and Apple® Safari®. For HVAC control system interfaces, Microsoft browsers such as Edge and Internet Explorer are commonly used.

Figure 4-2. The Java Control Panel is accessed through the Windows Control Panel. It can be used to determine the version of Java on the computer.

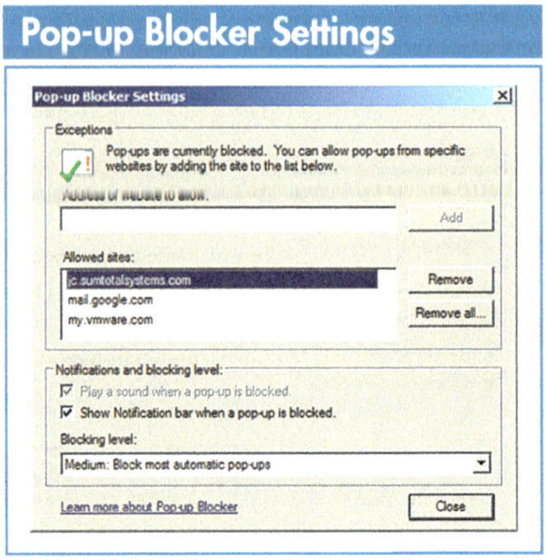

Figure 4-3. Pop-up blocker functions are accessed through the browser options.

Some HVAC control systems are viewable only on certain browsers. Also, browsers are available in different versions. Care must be taken to ensure that the correct browser and version are installed. Technical literature from the control system vendor should be consulted for details.

Some browser settings, such as security features and pop-up blockers, can affect a browser's operation as an HVAC control system interface. Pop-up blocking is a browser feature that prevents a website from opening new windows on its own. Pop-ups are normally used for marketing purposes, and most users find them annoying, so blocking them is usually desired. However, some websites, including some web-based supervisory controller interfaces, use pop-ups as part of their interface. Browser settings can be configured to allow pop-ups for certain websites only. **See Figure 4-3.**

The controller vendor's installation literature should be consulted regarding browser versions and settings. The facility IT staff must be consulted if their policies conflict with HVAC control system requirements. In some cases, exceptions may be allowed for a computer that is used only for the HVAC control system.

Windows Dependencies

To ensure that the HVAC control systems computer operates properly, there are a number of Windows-related applications that must be installed and functioning. These are known as dependencies. In other words, the system functionality is dependent upon these Windows functions. **See Figure 4-4.** Each dependency is frequently revised, and controller software may require a specific version in order to operate properly. Examples of common Windows dependencies include the following:

- Windows Installer: Windows Installer is a software utility that facilitates the installation, uninstallation, and maintenance of software.

- .NET Framework: The .NET Framework provides a software environment that can run applications written in a variety of programming languages. It also allows old and new applications to run on the same computer.
- Internet Information Services (IIS): Internet Information Services is a Windows web server. Specific features inside IIS may need to be turned on or off according to controller manufacturer requirements.
- Structured Query Language (SQL) software: SQL (commonly pronounced as "sequel") is a system for programming and managing data within a database. While mainly used with server class computers, it may also be needed with vendor programming tools as well.

Windows Dependencies

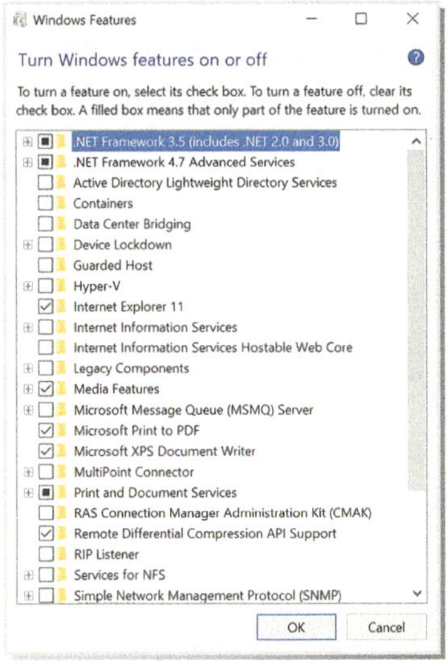

Figure 4-4. Windows dependencies are installed and removed through the Control Panel.

A facility's IT staff may replace a client computer with a new one as part of their standard upgrade cycle. The new computer may have a new OS and need to be completely reconfigured by the HVAC controls technician.

Utilities

A *utility* is a software tool that helps analyze, maintain, configure, or optimize the operation of a computer. Common utilities include antivirus software, backup software, disk-repair software, and encryption software. Utilities that are commonly used by HVAC controls technicians include system, network, and vendor utilities.

System utilities help optimize general system performance and provide security. Many system utilities are provided with the operating system, and some are available separately.

Network utilities are particularly useful for troubleshooting network performance problems. Some basic network utility features are included in the system utilities, but usually the IT staff has other, more sophisticated software. These utilities can typically view network traffic and diagnostics in real time. The software is loaded and run under the supervision of the IT staff.

Vendor utilities are configuration, backup, and diagnostic utilities that are provided by a controls vendor. Common features include an autodiscovery feature that automatically detects networked controllers, a notification feature that indicates when software updates are available, and diagnostic utilities that can be used if the controllers fail.

Peripheral Drivers

Any needed peripherals, particularly printers, must be installed and functioning correctly. **See Figure 4-5.** Printers are often used to print out HVAC control system information for viewing, sharing, and recordkeeping. If printers are changed or added without verifying proper operation, the customer may not be able to print alarms or other data. A computer may require specific software, known as a driver, to communicate with a peripheral. Drivers are always specific to particular device models and may need to be updated if other system software is updated.

Software Licenses

All software installed on the computer must have a valid license from the developer. Because this information may be required when requesting support, technicians should record license information and store it in a secure place.

NETWORK ACCESS

A client computer may access the HVAC control system network through a wired connection, wireless connection, or remote connection. Wireless and remote access

Printers and Peripherals

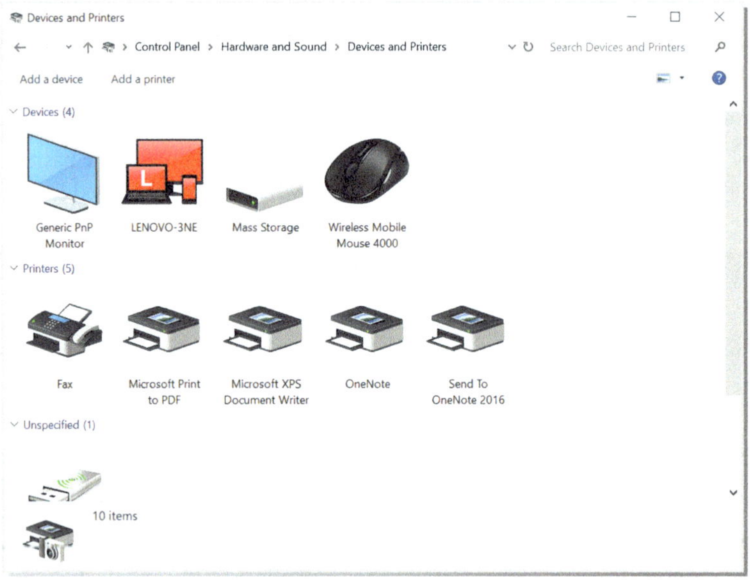

Figure 4-5. Printers and other peripherals are installed and configured through the Control Panel.

are very useful features, but add an additional layer of complexity, which may make troubleshooting more difficult.

Wired Access

A client computer with wired network access is connected to the network in a similar way as the web-based supervisory controllers and other nodes. The computer is connected via twisted-pair cable to a network switch. Sometimes the computer is at a workstation with a wall jack, which is wired to a switch in a telecom closet. Or, the computer (often a laptop) is connected directly to a network switch, which is either in the telecom closet or inside a supervisory controller enclosure.

Wireless Access

Wireless access is a feature that provides the ability to access an HVAC control system using the wireless capability of a computer. This allows an operator with a laptop to roam through a facility and access the control system without a network connection jack.

Wireless access is very common for facility networks, but technicians may need to verify with the IT staff that the HVAC control system is also wirelessly accessible. Also, there may be multiple wireless networks in a facility, and their access permissions may vary. For example, it is common for facilities to have a public wireless network with limited access, intended for guests, and a restricted network for employees. Wireless networks are commonly password-protected.

Remote Access

Remote access allows the control system to be accessed and monitored from locations outside of the facility's LAN, such as at other facilities, at home, or in transit (via a mobile application). One can receive alarms and control critical components at night or on weekends, which may help HVAC technicians avoid having to return to a facility to address a problem.

However, this kind of access may pose a security risk, as it makes it easier for unauthorized users to access the system. In order to provide a secure connection to locations outside of the facility's local network, the IT staff can set up a virtual private network (VPN). A *virtual private network (VPN)* is a software feature that provides an encrypted and secure method of remote access that mimics a local, physical connection to a LAN.

SYSTEM SECURITY

Security is a critical aspect of networked HVAC control systems. Unauthorized users could inadvertently change settings. Malicious users could disable or change the system operation in ways that negatively affect the facility and the occupants. If the HVAC controls system is linked with the facility's general computer network, company data and other operations may also be vulnerable. Router and OS firewalls are major parts of system security. **See Figure 4-6.** Passwords and antivirus software are also parts of a security plan.

Passwords

Passwords are used to restrict system access to certain users. HVAC controls technicians typically require high-level access, so it is important that they possess all necessary passwords. Technicians should communicate with the IT staff concerning the necessary passwords.

Firewall Settings

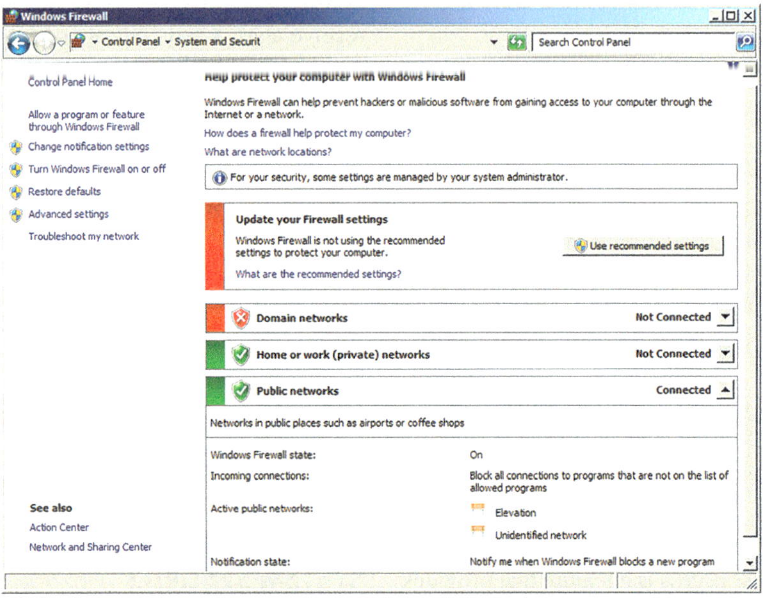

Figure 4-6. The Windows Firewall settings are accessed through the Control Panel. A firewall is used as part of a security plan.

Typically, a system administrator password is required, which allows the HVAC technician to load new programs or change program settings. This is the highest-level password on the network and allows full access to every device. If the IT staff is concerned about allowing this level of access, then the staff may grant a temporary system administrator password, which will allow a technician to load software and configure the system. The IT staff would then change or disable the password after the technician has finished.

Antivirus Software

Nearly all computers, especially those connected to the Internet, are loaded with antivirus and other security software. Antivirus software helps prevent the execution of hidden, malicious application programs that corrupt and/or upload data to other servers. Antivirus software must be updated periodically to keep up with new virus threats.

Due to its tight controls, antivirus software commonly interferes inadvertently with the operation of desired software, such as HVAC control system software. Technicians must be careful to check antivirus software settings and work with the IT staff to avoid this problem.

CONTROLS SOFTWARE

Specific HVAC control system software may be provided by the controls vendor in order to program the web-based supervisory controllers or field bus control points. Care should be taken as it may be necessary to load specific OS components for these software packages. Also, any updates to the computer's system software, either intentional or accidental, may interfere with the proper operation of the control system software.

Basic Software Troubleshooting

The vast majority of HVAC control system problems are due to issues with system software. While there are literally thousands of possible problems, a systematic and general troubleshooting approach is also possible. Once a problem has been reported, the first step is to ask the following basic questions:

1. When did the problem start?
2. Did the computer operate properly before that time?
3. Has anyone, such as a member of the IT staff, made any recent changes to the system? (If so, what changes have been made specifically?)
4. Have any upgrades been performed recently? (Again, if so, what specific upgrades have been performed?)
5. Has any additional software been loaded or removed? (If so, what software has been loaded or removed, when, and by whom?)

The answers to these questions may help determine the problem with the software. For software, the most basic troubleshooting tool is the Windows Control Panel, which contains many system software options.

Review Questions

1. What are the three types of computer configurations for accessing HVAC control systems?
2. What are the primary reasons for common IT policies?
3. What is the difference between service packs and patches?
4. What are the most common web browsers used to access HVAC control systems?
5. What are some examples of utilities?
6. What are three ways client computers access the HVAC control system network?
7. What are the capabilities of a system administrator password?
8. What causes the vast majority of HVAC control system problems?

Servers 5

One of the most important devices in an HVAC control system network is the server, and it can also be one of the most difficult devices to install, maintain, and troubleshoot. The server is also usually the most expensive device in the system and requires the most cooperation and communication with the facility IT staff. A failure of the server is considered a major or even catastrophic failure.

HVAC CONTROL SYSTEM SERVERS

A *server* is a dedicated computer that provides application services to other computers on a network. The other computers are called clients. There are many different hardware variations and applications for servers. An HVAC control system server provides functions that controller-level software cannot perform. The server typically provides scalability, multiuser access, centralized functions, protocol translation, data storage, and data analysis and publishing over the local network and/or the Internet.

Server Size

Not all networked HVAC control systems require a server. Small buildings may only have one or two web-based supervisory controllers, which would not require a server. However, when the system is over a certain capacity, a server must be installed. This capacity varies between manufacturers, but the common range is four to six web-based supervisory controllers.

Most manufacturers provide different sizes or levels of servers. Small-capacity servers are appropriate for systems with a small number of web-based supervisory controllers and limited data storage. Large-capacity servers are used for systems with a larger number of web-based supervisory controllers and much greater data storage needs. There are many differences between small and large servers, primarily related to the sophistication and cost of the server computer itself. **See Figure 5-1.**

Server Capabilities

Regardless of system size, the customer may want or need the unique capabilities provided by a server. Unfortunately, it is common for customers to be disappointed in the limited capabilities of a system that was deemed too small to require a server. An installation technician should review a system's specifications with the customer to ensure that all desired capabilities are addressed. While each server implementation may vary, common capabilities of servers include data storage and retrieval, graphics, passwords, and tools.

Data Storage and Retrieval. The most important capability a server provides is a large amount of raw data storage space and the ability to access the data in meaningful ways. The server stores data in an SQL database, which uses a standard industry format. There are three general categories of data: trend data, alarms, and audit trails.

Comparison of HVAC Control System Server Sizes		
Feature	**Small Servers**	**Large Servers**
Number of controllers supported	Typically 6 or fewer	Typically up to 100
Number of simultaneous users supported	Typically 6 or fewer	Typically up to 25
Data storage	Limited	Thousands of points; years of data
Operating system	Standard version	Server version
Hardware required	Standard; less expensive	Server-class; more expensive
Difficulty of setup	Easier	More difficult; more prerequisites

Figure 5-1. HVAC control system servers are generally classified as either small or large. The features of the two vary greatly.

Trend data is a large set of values for system variables, such as for temperature, humidity, pressure, and other variables, that are recorded at regular intervals. This data is analyzed for trends, or changes over time, that may indicate equipment failure, energy efficiency issues, and other conditions. **See Figure 5-2.**

An *alarm* is a notification that the value of a system variable has reached a preset limit. An alarm notifies the building staff of the unusual condition, which may indicate a problem. **See Figure 5-3.** The server maintains the variable limit points, records of alarm events, and alarm-handling instructions, such as sending an email.

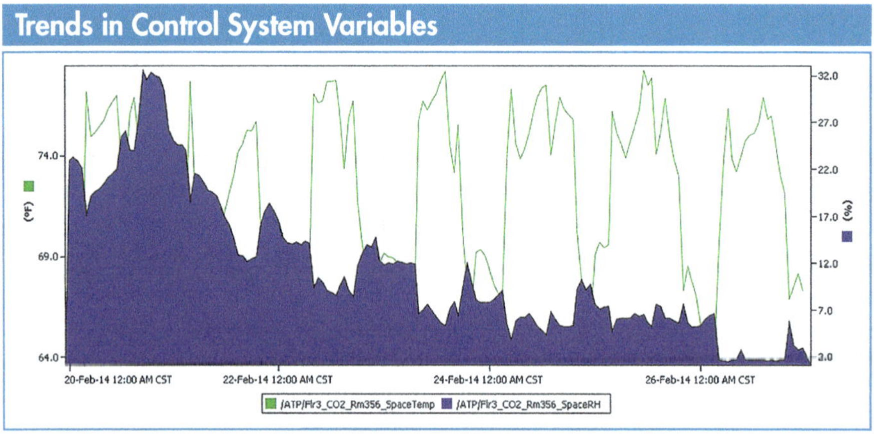

Figure 5-2. Trends are relatively gradual changes in control system variables over time. Trends can reflect normal system changes or indicate developing problems.

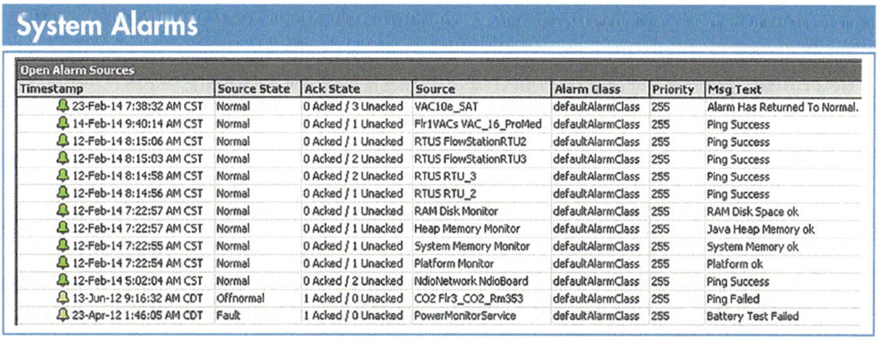

System Alarms

Open Alarm Sources

Timestamp	Source State	Ack State	Source	Alarm Class	Priority	Msg Text
23-Feb-14 7:38:32 AM CST	Normal	0 Acked / 3 Unacked	VAC10e_SAT	defaultAlarmClass	255	Alarm Has Returned To Normal.
14-Feb-14 9:40:14 AM CST	Normal	0 Acked / 1 Unacked	Flr1VACs VAC_16_ProMed	defaultAlarmClass	255	Ping Success
12-Feb-14 8:15:06 AM CST	Normal	0 Acked / 1 Unacked	RTUS FlowStationRTU2	defaultAlarmClass	255	Ping Success
12-Feb-14 8:15:03 AM CST	Normal	0 Acked / 2 Unacked	RTUS FlowStationRTU3	defaultAlarmClass	255	Ping Success
12-Feb-14 8:14:58 AM CST	Normal	0 Acked / 2 Unacked	RTUS RTU_3	defaultAlarmClass	255	Ping Success
12-Feb-14 8:14:56 AM CST	Normal	0 Acked / 1 Unacked	RTUS RTU_2	defaultAlarmClass	255	Ping Success
12-Feb-14 7:22:57 AM CST	Normal	0 Acked / 1 Unacked	RAM Disk Monitor	defaultAlarmClass	255	RAM Disk Space ok
12-Feb-14 7:22:57 AM CST	Normal	0 Acked / 1 Unacked	Heap Memory Monitor	defaultAlarmClass	255	Java Heap Memory ok
12-Feb-14 7:22:55 AM CST	Normal	0 Acked / 1 Unacked	System Memory Monitor	defaultAlarmClass	255	System Memory ok
12-Feb-14 7:22:54 AM CST	Normal	0 Acked / 1 Unacked	Platform Monitor	defaultAlarmClass	255	Platform ok
12-Feb-14 5:02:04 AM CST	Normal	0 Acked / 2 Unacked	NdioNetwork NdioBoard	defaultAlarmClass	255	Ping Success
13-Jun-12 9:16:32 AM CDT	Offnormal	1 Acked / 0 Unacked	CO2 Flr3_CO2_Rm353	defaultAlarmClass	255	Ping Failed
23-Apr-12 1:46:05 AM CDT	Fault	1 Acked / 0 Unacked	PowerMonitorService	defaultAlarmClass	255	Battery Test Failed

Figure 5-3. System alarms are tripped when a control point value falls outside of its expected range.

SPECIAL DATA STORAGE REQUIREMENTS

The data storage capability of a server is critical for certain HVAC systems. For example, in pharmaceutical facilities, the operator actions contained in audit trails are legally controlled and mandated by federal regulations. Also, regulatory agencies may require this data on very short notice. Hospitals are another type of facility that requires a large amount of HVAC system data storage, particularly for operating rooms, drug storage refrigerators, tissue sample freezers, and infectious disease isolation rooms.

An *audit trail* is a record of all the changes made to system parameters and settings, including the date and time of each change as well as the user identification of who made the change. For example, an audit trail can show who discarded an alarm and when, who put a valve or damper in override, or who changed the temperature setpoints for an HVAC unit. This information is useful for analyzing staff functions, training needs, and problems.

Each web-based supervisory controller typically also has on-board data storage, but it is limited to a range of about a day's worth of data to about a month's worth of data. With a greater amount of data storage available, servers provide the ability to store and review data recorded over a year ago or longer.

Graphics. Most modern web-based HVAC control systems feature a graphics package, which includes pictorial views of the building and HVAC equipment. **See Figure 5-4.** This is in addition to the standard textual representation of the building information. Each system differs, with some using graphics only while others use a combination of graphics and text. Graphics often require a large amount of data storage space, so they are stored on and accessed from the server.

Passwords. A password scheme is typically implemented at both the server and the web-based supervisory controllers. A *password scheme* is a comprehensive plan for safeguarding user access levels, such as limiting who has access to which inputs, outputs, devices, and functions. Time-of-day scheduling may also be used to determine when a password is active. Password schemes may be very simple or very complex, with most manufacturers using IT standard network-based password and security schemes. This includes the single sign-on (SSO) capability to allow a single password for the computer and any dependent applications. The HVAC controls technician may need an administrator-level password in order to access the server and its functions.

Tools. Controller vendors typically have a number of software tools, and the server is a good place to load them. The tools typically include programming and diagnostic tools for the web-based controllers. These tools can be used by the HVAC controls technician or IT staff as needed. In addition, the server's operating system (OS) includes diagnostic tools to help troubleshoot OS-related problems. Also, third party tools can provide many additional options. It is strongly recommended that HVAC technicians install as many of these tools as possible during the server software installation, as these tools can be useful when troubleshooting a problem.

System Graphics

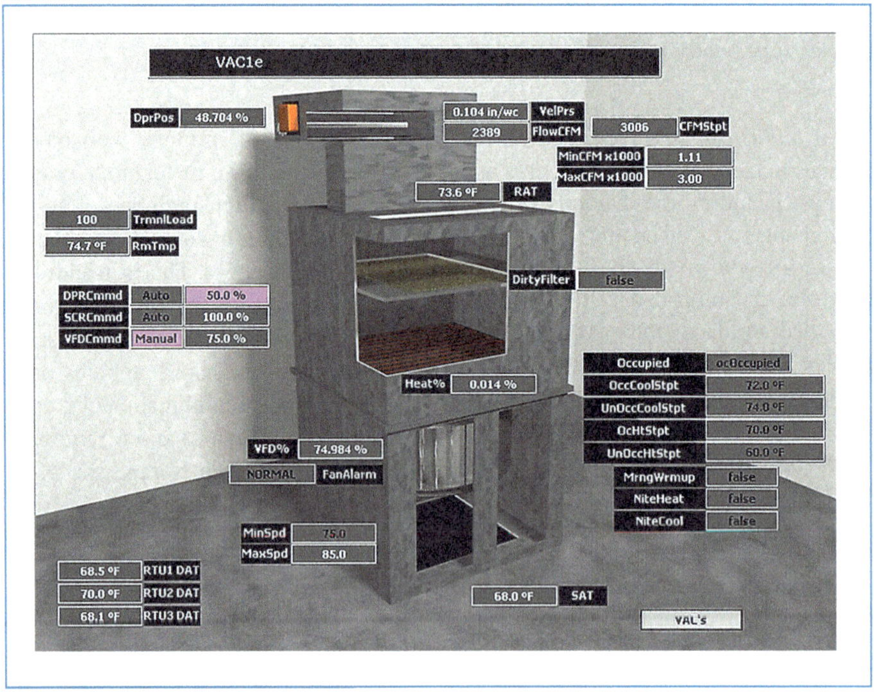

Figure 5-4. System graphics help visualize the association between control point data and the HVAC equipment.

Large servers are typically installed in a rack mount server cabinet. These cabinets are standardized for IT equipment and use special mounting brackets to stack the devices vertically.

Server Purchase

One of the first major decisions involves the purchase of the server, which may be by either the HVAC controls vendor or the facility IT department. Large-capacity servers are usually purchased by the IT staff, and small-capacity servers are often purchased by the HVAC controls vendor. If the IT staff purchases the server, then any prerequisites must be provided for the IT department beforehand. It is also possible to purchase "turnkey" servers that are already loaded and tested at the HVAC control systems factory. These servers save a lot of time in the field. In any case, the cost of any server as well as the software and the technician's time to configure it should be estimated and included in a quote. Each server installation will be somewhat different.

Physical vs. Virtual Servers

The most common type of HVAC control system server involves a dedicated, server-class computer hardware platform. However, there are several disadvantages to this setup. Some facilities have multiple building systems that require server software, such as telecom systems, fire and life safety systems, security systems, closed-circuit TV systems, and others. If each system required a dedicated physical server, all the hardware would involve high energy costs to operate, high energy costs to cool, and high maintenance costs. Also, many of these server computers only use a small percentage of their computing and storage capabilities, which is highly inefficient.

Many IT departments are converting their HVAC and other systems from operating on a hardware platform to running in a virtual server environment. A *virtual server* is a software emulation of a complete server operating system, separate and independent from other virtual servers on the same hardware. A single, high-capacity server computer can be used to run multiple virtual servers. Each virtual server is designated a separate OS and application software. This reduces the number of physical servers, which significantly increases computing efficiency and decreases energy costs. It can also reduce server maintenance costs as well. In almost all cases, the facility IT staff is responsible for implementing the server virtualization software. The HVAC controls technician then loads the HVAC controls software on the virtual server after it is in place.

> A virtual server operates as if it has dedicated use of the physical resources on the server computer (such as RAM, CPU time, and port access), though the resources are actually being shared with other virtual servers on the same physical machine.

Server Locations

Servers are typically located in either an HVAC maintenance office, which is more common for small servers, or in a telecom closet, which is more common for large servers. **See Figure 5-5.** The main advantage of a maintenance office location is that it allows easy access for an HVAC controls technician. However, this is also a disadvantage because access is also easier for unauthorized users, which can cause problems. The main disadvantage of a telecom closet is the lack of easy access, as IT staff controls this access and perhaps even requires an escort during a visit. This also means that IT staff have to provide space for the server and physical services such as power and network connections.

Large Servers

Figure 5-5. Large servers are usually located in a rack mount cabinet in the telecom closet.

SERVER SOFTWARE

There are many similarities and some major differences between the setup of a server and the setup of a client computer for a facility. Much of the client computer setup, software, and configuration is applicable, but a server also requires additional considerations.

Operating Systems

The single most important software consideration for a server is its operating system. It is absolutely vital that the OS be compatible with any vendor software needed. Small servers may use a standard client computer OS, such as a version of Microsoft Windows. Large servers utilize a version of Windows Server OS, though some are loaded with Linux. There have been many versions of Windows Server, including Windows Server 2003, 2008, 2008R2, 2012, 2016, and 2019. Because building operators and HVAC controls technicians may encounter some of the older versions at existing installations, they must be able to support these systems.

New control system installations will use a recent version of Windows Server, but the latest version may not be the best choice. Software that runs on the server, including control system software, must be thoroughly tested by the vendor on that platform before being released. Therefore, certain software may be unavailable for a server with the latest OS. All necessary software and their requirements must be researched before making server purchase decisions. Software vendors publish any OS requirements in the installation documentation.

Each of these server OSs have different editions available. For example, some

Windows Server software is available in 10 different versions. It is important that the appropriate version be loaded on the server computer for proper operation and support.

32-Bit and 64-Bit Architectures. Operating systems and computers have either 32-bit or 64-bit architectures. Older computer technology mostly involves 32-bit computers and OSs. The "32-bit" and "64-bit" designations refer to the size of each unit of information that can be processed by the CPU. These architectures also limit how much RAM a computer can use. Today, 64-bit computers and OSs are affordable and practical, and many IT departments use this architecture with both their servers and their client computers. Computers with 64-bit architectures can handle large data collections more efficiently and utilize more RAM than 32-bit computers, which results in faster application speeds.

HVAC controls technicians need to be aware of the distinction between 32-bit and 64-bit architectures as this distinction can dramatically affect the control system operation. However, the choice is often reserved for the facility IT staff. If the facility decides to use 64-bit computers, this typically involves both hardware and software upgrades. It may be a good opportunity to switch to a virtual server environment. Also, while some 32-bit applications may run on a 64-bit computer, others may not, particularly programming software for legacy (outdated technology) controllers. The HVAC controls technician should research all possible software conflicts and work with the IT staff to find solutions. One possible solution is to retain a 32-bit laptop for use with older software.

Server Service Packs and Patches. As with client computers, service packs (SP) and patches may be available for servers. An SP or patch must be installed with caution, as it may or may not be compatible with the HVAC control system software.

Server Utilities

In addition to any general system, network, and vendor utilities, which are also loaded on client computers, servers may require several additional specific utilities. These are software applications with features directly related to the services provided by the server.

The specific types of server utilities recommended for different systems vary, but a common type involves the use and management of SQL databases. **See Figure 5-6.** One or more software applications may be used for database configuration, management, diagnostics, backup, and data retrieval. SQL data retrieval utilities are used for extracting data from an SQL database and saving it in a spreadsheet or other document format. This software is typically provided by the HVAC controls vendor.

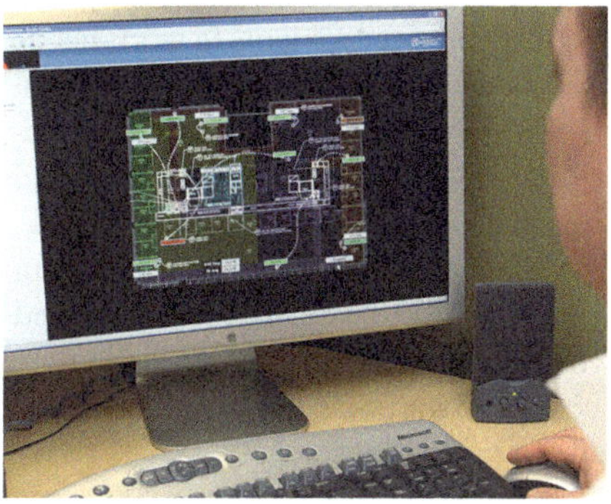

Servers gather information from all the networked controllers, and users can access this information in various formats from a client computer.

SQL Utility Software

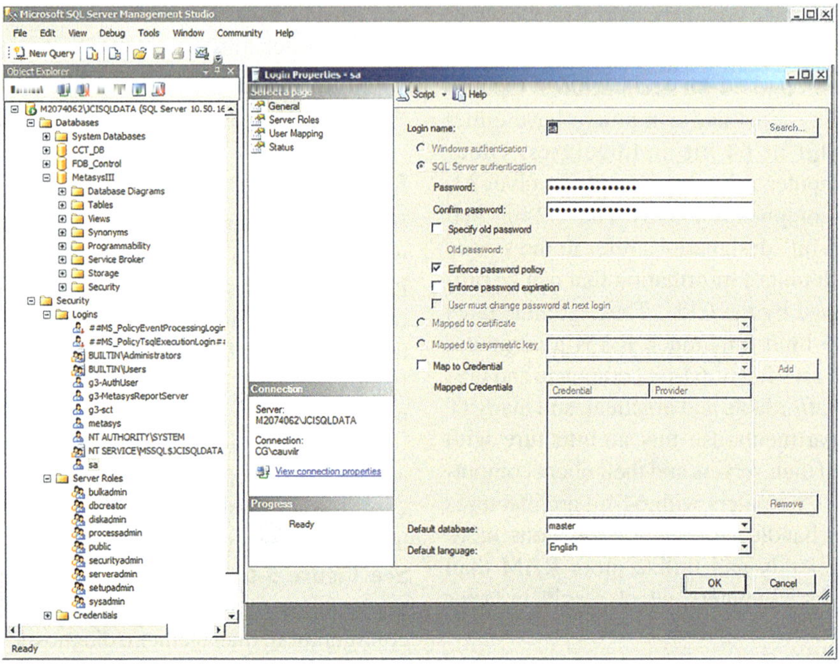

Figure 5-6. SQL utilities are used for configuration and troubleshooting.

Remote Access Software

An HVAC control system server is typically in the telecom closet, so physical access can be a problem. If work needs to be performed on a server, this can often be done via remote desktop software, which allows one computer to view and control another computer over a network. There are many third-party solutions, but Remote Desktop Services is included with Windows. **See Figure 5-7.** Remote access software has to be setup and enabled properly.

If a server must be accessed from outside the facility, the IT staff can often set up a VPN, which provides a secure connection for remote access applications. This is helpful for remotely logging into and troubleshooting servers.

Other Software

As with a client computer, a server requires a number of ancillary software programs to support the operating system and the HVAC control software. The considerations are generally the same, with the prime concern being compatibility between the versions of different software programs that interact with each other. The HVAC control system software requirements should always be checked closely for potential compatibility issues. Types of additional software typically loaded onto a server include the following:

- Java
- web browsers
- Windows dependencies
- peripheral drivers
- antivirus software
- control system software

Remote Desktop Software

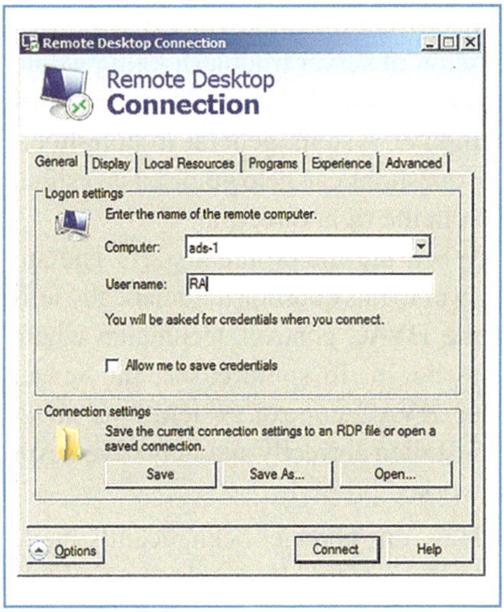

Figure 5-7. Remote Desktop Services is a Windows feature that can be used for remote access and troubleshooting.

Client Access Licenses

Windows Server OSs require a client access license (CAL) for all client computers that access the server. This legally limits the number of simultaneous connections to the server. The server OS documentation, as well as the vendor HVAC software, will indicate how many simultaneous connections are available. Typical numbers are 5, 6, 12, or 25 simultaneous connections.

SERVER SETUP AND MAINTENANCE

While the server is installed by the facility's IT staff, the HVAC technician is responsible for the installation of the vendor controls software, ongoing maintenance, and troubleshooting. This means that the technician must be familiar with the specific vendor software loading procedures and be able to follow complex installation instructions, which may be hundreds of pages long. While each installation process is different, the general steps are as follows:

1. Qualify the server to be sure it meets the requirements.
2. Install the physical components of the system into the appropriate location and enclosure, such as a rack mount server cabinet.
3. Start up the server and log on.
4. Verify the Windows Server OS and the Windows dependencies.
5. Install the HVAC control system software.
6. Start up the system and conduct a test run.
7. Correct any problems found.

Technicians must also be able to access OS and vendor software updates and resources on the Internet. Vendors also usually provide a technical support phone number for installation, maintenance, and troubleshooting.

A server upgrade is different from a new installation. The server computer and OS remain the same, but the control system software is upgraded (assuming the new software versions are compatible). Upgrading a server is usually faster and easier than installing a new one, but the server will be unavailable during the process. The customer must be notified beforehand as to when a server will be down and for how long. Also, all server data should be backed up in case of problems.

Server Maintenance

Server maintenance generally consists of periodically running diagnostic software to ensure that the server data is intact, that there are no computer viruses, and that the hardware is functional. The HVAC control technician and the IT staff must establish who has this responsibility. Failure to do this can lead to confusion in the event of a failure and may delay repairs.

Server Backup and Imaging

The data on the server must be backed up at regular intervals and also before any major work or upgrades are performed. Monthly or weekly backups are common. The backup is often performed by the IT department using network software that runs on a set schedule. Again, responsibilities must be discussed beforehand to avoid misunderstandings and problems.

Some problems, such as computer viruses, can corrupt data across an entire server. For this reason, it is recommended to image the server's hard drive periodically as well. Imaging software makes an exact copy of an entire hard drive as one file, called an image. If needed, the imaging software can then use this file to restore the entire server back to its previous state. It is recommended to keep images for multiple intervals into the past, in case a problem is not detected right away and a corrupted hard drive is imaged. This is a more complex procedure than a backup procedure, however, so backups are still useful for problems limited to control data.

BASIC SERVER TROUBLESHOOTING

Due to the large number of hardware/OS/software combinations, an extensive discussion of server troubleshooting would not be practical. However, as with any computer system, general troubleshooting questions can help guide an investigation in the right direction:

- When did the problem start? The answer to this question automatically tells the HVAC controls technician where to begin. In some cases, the server or HVAC control system was never installed correctly and may have never worked properly.
- Have any changes been recently made to the system? If so, what were they? Often, the technician finds that software was recently added, removed, or changed, which can lead to the source of the problem. Again, this is an ideal time to speak with the facility IT staff.
- How old is the server or HVAC control system? An older system may need patches and upgrades that have not been performed. Also, the Windows components may no longer be supported or available.

CLOUD COMPUTING

Cloud computing and storage allows commercial networks to utilize shared system resources that are located off site and accessed via the Internet. This technology has been made feasible by low-cost storage and computers, as well as greater access to high-speed networks. With this arrangement, a commercial facility does not have to own or maintain extensive computer resources, which may also reduce the size and complexity of the necessary on-site network. Cloud-based resources are purchased based on the amount of data storage and applications needed. For building-control services, the information stored in the cloud may then be analyzed by a third party to identify energy waste and other building conditions.

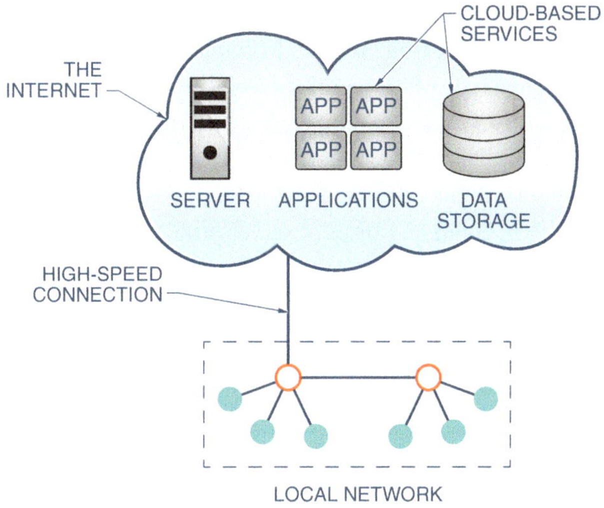

Most problems are either Ethernet network problems or software problems and have a relatively small number of causes. Common Ethernet network problems include the following:

- disrupted communication due to failed, disconnected, or moved cable at a router
- disrupted communication due to changed IP address
- disrupted communication due to changed infrastructure (typically a router)
- blocked communication due to network software changes, such as changed security settings

Common software problems include the following:

- unstable OS due to Windows Server updates
- disrupted communication due to changed server configurations (often inadvertently)
- malfunctioning HVAC controls vendor software
- blocked communication due to security or firewall changes
- blocked communication due to antivirus software
- computer viruses infecting server or clients

Review Questions

1. What determines whether a server is needed for an HVAC control system?
2. How are servers used to analyze trend data?
3. What is an audit trail?
4. Who typically purchases an HVAC control system server?
5. What are some advantages of a virtual server?
6. What is the main advantage and main disadvantage of locating a server in an HVAC maintenance office?
7. What do "32-bit" and "64-bit" refer to in computer architectures?
8. What is a common type of server utility software?
9. How are server setup responsibilities typically divided between IT staff and HVAC controls technicians?
10. When should server data be backed up?
11. What are the most common causes of server problems?

Wireless Networking 6

One of the most exciting developments in HVAC control networks is the widespread use of wireless networking technology. This technology is now widely used in both residential and commercial buildings. Almost all commercial buildings today have some type of wireless network infrastructure. With wireless networking, desktop and laptop computers can establish high-speed connections with a facility LAN without physical cables.

WIRELESS NETWORKING

Wireless networking is the use of radio frequencies as a communication medium for computer networks. Wireless technology is also known as Wi-Fi® or WiFi. The use of wireless networks has expanded greatly in recent years, though primarily for noncritical applications. The primary portion of the HVAC control system, which is operating the equipment, is typically a wired network. However, wireless networking technology is commonly used by operators and technicians to access the system using client computers, smartphones, and tablets. Many web-based supervisory controllers even have built-in wireless adapters from the factory. **See Figure 6-1.** HVAC controls technicians should be familiar with the basic operation and common problems of wireless networks.

For security reasons, there are normally two types of wireless networks available in a facility: public networks for access to the external Internet only and restricted networks for internal use. Public networks may not be password-protected, but internal networks are.

Wireless networking uses the 2.4 GHz and 5 GHz radio frequencies. Other devices that may emit these frequencies include microwave ovens, television satellites, medical equipment, fluorescent lamps, cordless phones, baby monitors, and radio-controlled aircraft.

Wireless Networking

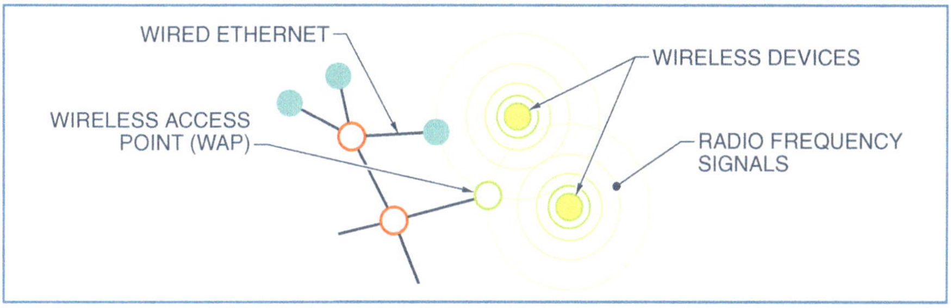

Figure 6-1. Wireless networking allows devices to connect to each other and/or to a wired LAN using radio-frequency signals.

Wireless Networking Advantages

There are a number of advantages to wireless networks. Most importantly, a wireless network allows users with laptop computers to access the system from anywhere within the wireless coverage area. This can be extremely helpful to HVAC technicians who may need to troubleshoot a problem by inspecting HVAC equipment while also reviewing control system data. With a laptop and control system access via a wireless network, the technician can work nearly anywhere within the coverage area. **See Figure 6-2.**

Using Wireless Networks

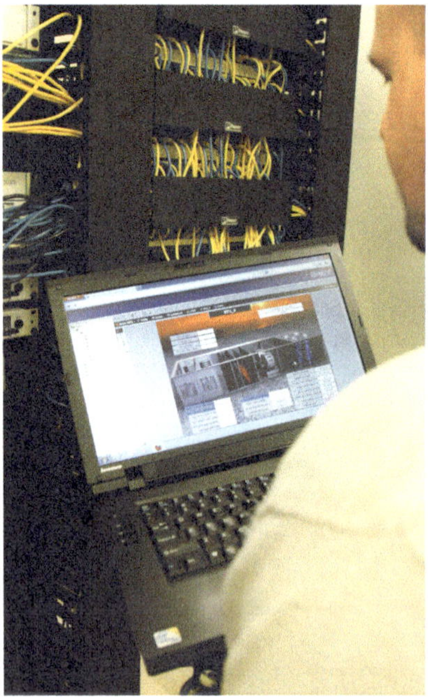

Figure 6-2. Wireless networks allow HVAC controls technicians to use laptop computers anywhere necessary when troubleshooting.

Also, the use of wireless infrastructure reduces the need for Ethernet media such as cables and jacks. Since cables are typically installed inside walls and ceilings, expanding a network after a building is constructed can be difficult. However, a wireless network provides expandability with a relatively low impact on the building.

Third, wireless networks are relatively simple to implement. Adding the necessary network hardware is fairly straightforward, and all new client computers and devices already include wireless capabilities.

Finally, wireless networking extends control system access to mobile devices, which otherwise would not be able to access the network. Tablet computers and smartphones can now be used to view real time control system data.

Smartphones are an increasingly common way to access HVAC control system information.

Wireless Networking Disadvantages

Wireless networking also involves a number of disadvantages, which have kept it from being implemented in certain applications. Even when successfully employed, however, certain additional measures may be required to maintain the integrity of the overall network.

WIRELESS RESIDENTIAL THERMOSTATS

One of the important outcomes of wireless technology is the development of networked thermostats for residential and light commercial HVAC systems. Residential and light commercial buildings were slow to adopt network-based control because there usually was no easy way to run physical cables from the building LAN to the thermostat location. However, with wireless networks, interfacing between the HVAC system and a LAN is much easier.

Some manufacturers even provide means to configure a connection to an HVAC thermostat from outside a LAN using the Internet. Because residences typically do not have fixed IP addresses, this feature usually requires a special IP-forwarding service. This service allows users to monitor and control their home temperature even when they are away from home.

Wireless thermostats are available for multistage heating/cooling systems, heat pumps, and hydronic heating systems. Installing a wireless thermostat is similar to any other thermostat, with the added process of configuring a network device. The general steps to install and configure a wireless thermostat include the following:

1. Turn off the power to the HVAC unit.
2. Remove the old thermostat and baseplate, and label the wires as needed.
3. Mount the new thermostat and attach the wires to the proper terminals. Install a battery if necessary.
4. Turn on the power to the HVAC unit. The thermostat should start operating.
5. Using instructions from the manufacturer, log in to the thermostat using a smartphone, tablet, or computer, and set up the wireless network and other parameters.

Note: If multiple wireless thermostats are present on the same network, they must be individually configured. Each has a unique MAC address from the manufacturer.

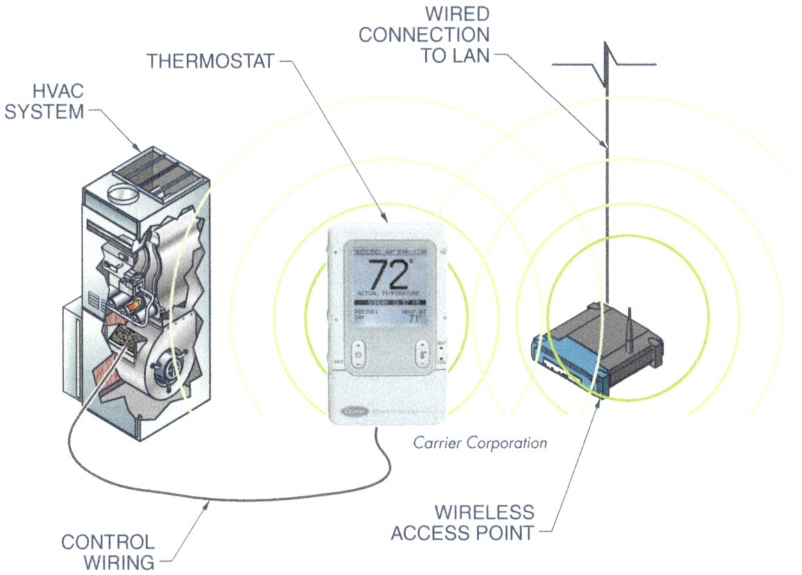

Carrier Corporation

Security is a primary concern. Wireless signals can broadcast up to a distance of about 300 ft, which could reach areas outside of a facility's property where unauthorized users could access the system. For this reason, wireless networks are prohibited at some critical facilities, such as military and certain industrial facilities. To address this issue, wireless networks employ a number of security technologies, including passwords. Passwords are encrypted and verified through a number of different protocols, including WEP, WPA, and WPA2. **See Figure 6-3.** Firewall software can also provide security between a LAN and the Internet.

Wireless Password Prompts

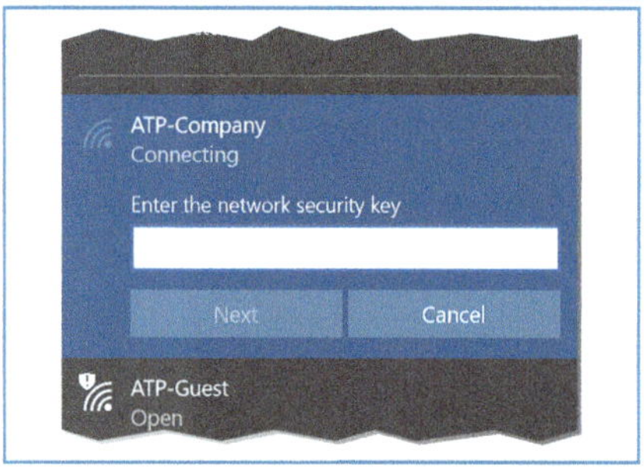

Figure 6-3. Wireless network security requires encrypted passwords in order to connect to a network.

Since it uses radio-frequency signals for transmission, wireless networking is vulnerable to interference from equipment that emits radio-frequency signals. Interference may be caused by many devices, such as radio equipment and mircowaves. Interference causes connection problems and data corruption. This may be merely inconvenient or it may be dangerous. In facilities with certain critical equipment, such as medical scanners or sensitive security systems, wireless networks are prohibited due to these interference concerns.

Radio-frequency signals can penetrate many types of building materials, but these signals are attenuated (reduced in strength). Different materials and thicknesses affect the amount of attenuation and create an uneven signal strength throughout a facility. This attenuation along with radio interference can cause a strong signal to disappear within only a few feet. In order to establish good wireless signal coverage over an area, wireless devices must be installed at strategic locations, such as in conference rooms and central offices.

Wireless Access Points (WAPs)

A wireless network is created with wireless access points (WAPs). A *wireless access point (WAP)* is a network device that is connected to a wired LAN and provides access to the LAN by wireless nodes via radio-frequency signals. **See Figure 6-4.** WAPs may also be called just "access points."

For very small wireless networks, such as residential installations, the WAP functionality is usually integrated into a router, which is then known as a wireless router. For larger areas, such as commercial or industrial facilities, many WAPs are needed to provide uniform wireless coverage. These devices are typically WAPs only and are wired back to a switch or router. These WAPs are installed in strategic locations throughout a facility.

WAPs are initially set up using a browser and are accessed via a common default IP address, such as 192.168.1.1.

After the password and other parameters are configured, the IP address may be changed as needed.

Wireless Access Points (WAPs)

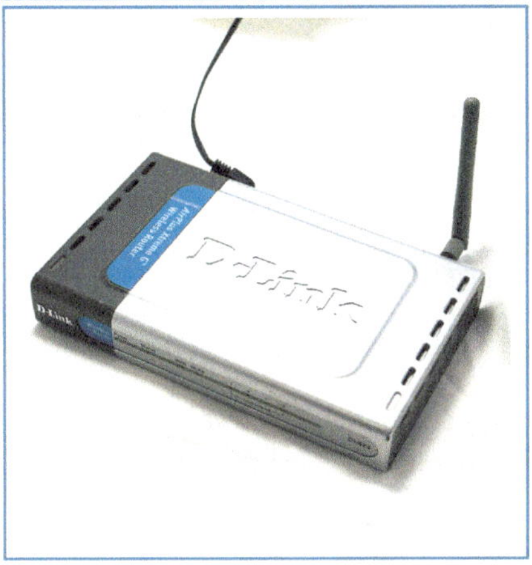

Figure 6-4. Wireless access points (WAPs) link the wired and wireless portions of a network.

Wireless Network Adapters

In order to use a wireless network, each computer and device needs a wireless network adapter. A *wireless network adapter* is a component that transmits and receives radio-frequency signals for a network node. Modern laptops have built-in adapters, while some old laptops may need a wireless adapter card. Desktop computers may have a built-in adapter or can use one that plugs into a USB port. Driver software is provided by the manufacturer and must be loaded in order for an adapter to work properly. Smartphones and tablets also have built-in wireless capability.

A wireless node is labeled with the designation for the applicable wireless technology standard(s).

Wireless Technology Standards

Like other communication protocols, wireless networking technology is defined by standards, the most important being the Institute of Electrical and Electronics Engineers (IEEE) standard IEEE 802.11. This standard is divided into sections for various related technologies, each designated with lowercase letters. The specific technology described in a particular section is therefore identified by the standard number and section letter, such as 802.11a.

The 802.11 standards range from 802.11a to 802.11ai (as of 2018), with more in development. Each standard defines radio frequency, bandwidths, modulation, and other radio parameters in order to optimize the performance for certain applications. **See Figure 6-5.** HVAC controls technicians should be familiar with these designations because WAPs and wireless adapters are specified by the wireless standards they are compatible with, and the two designations must match. Most wireless adapters and WAPs are compatible with multiple technologies, which are indicated in their designations, such as 802.11b/g/n.

Comparison of Common Wireless Technology Standards

Standard	Year	Frequency*	Maximum Data Rate†	Approximate Range‡	Notes
802.11b	1999	2.4	11	35	Prone to radio interference from other 2.4 GHz electronics
802.11g	2003	2.4	54	35	Prone to radio interference from other 2.4 GHz electronics and other 802.11g nodes on overlapping channels
802.11n	2009	2.4 5	600	70	Maximizes data rate by choosing optimal frequency, antennae, and channels for the environment
802.11ac	2013	5	1300	70	"Wave 1" in 2013 under draft standard; partial implementation
			2340		"Wave 2" in 2016 includes more technologies for better performance

* *in GHz*
† *in Mbps*
‡ *in m, for high data rate signal*
Note: Range is highly dependent on output power, obstructions, and interference from other radio sources. Range can also be extended for lower data rates.

Figure 6-5. There are multiple wireless technology standards, which vary in specifications and capabilities. All wireless devices must be compatible with the same standard(s) in order to communicate with each other.

Wireless Network Utilities

Each wireless device has utilities that are used to set up and examine the wireless networking configuration and capabilities. **See Figure 6-6.** These utilities typically include the following functions:

- turning the wireless adapter on and off, if available through software
- identifying all available wireless networks
- connecting with a specific network
- entering a password for a wireless network if needed
- checking signal strength as the device is moved

> Wireless technology standards define parameters in Layer 1 (physical) and Layer 2 (data link) of the OSI model.

WIRELESS NETWORK TROUBLESHOOTING

Many tools are available for troubleshooting wireless networks, including both handheld hardware and software. Features and capabilities vary, but several analysis functions are common. Most provide a real-time performance summary on the overall health of the wireless network. This includes interference detection and analysis and visibility of all wireless traffic.

Wireless problems can often be intermittent. Therefore, it is common to use diagnostic tools to monitor a wireless network for extended periods. Alarms can be set to alert a technician to wireless network security and performance issues. Features like these are important for troubleshooting wireless network problems.

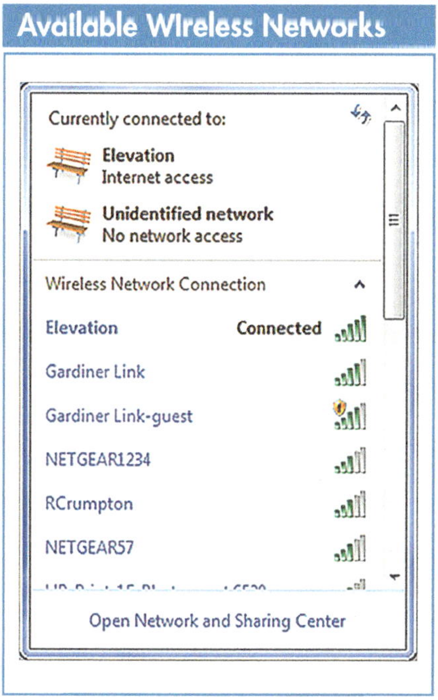

Figure 6-6. A control panel in Windows displays all of the currently detectable wireless networks, including their relative signal strength.

Significant problems with a wireless network should be addressed by the facility IT staff. However, several simple checks can be performed by HVAC controls technicians to solve many simple and common problems. Some of these simple checks include the following:

- Ensure that the power supplies for the WAP and other devices are turned on and plugged in.
- Ensure that all cables are connected securely.
- Reboot the computer or WAP. Disconnect the power supply for at least two minutes and then reconnect.
- Relocate a device to receive a stronger signal. Many software utilities display signal strength information. Sometimes moving a device only a few feet can significantly improve signal strength. If relocating a device is not an option, consider adding another WAP or a repeater, which boosts signal strength.
- Verify which network is active. Multiple wireless networks are often available in a commercial building. The device may be connected to the wrong network. Wireless software utilities can be used to change which network is being used.
- Check if a password has been changed or has been entered incorrectly.
- Ensure that the web-based supervisory controller is operational.

Wireless equipment relies on antennae for sending and receiving radio signals. Sometimes antennae are inside cases and not easily seen, but most are external. Equipment manuals provide instructions about adjusting antenna orientation.

Review Questions

1. Why may there be two separate wireless networks in a facility?
2. What are some advantages of wireless networks for HVAC control systems?
3. What are some disadvantages of wireless networks?
4. What hardware is needed to create and use a wireless network?
5. What is the primary standard for defining wireless networking signaling?
6. What functions are typically provided by wireless network utilities?
7. How can technicians troubleshoot intermittent wireless network problems?

Diagnostic and Troubleshooting Tools 7

An HVAC controls technician must be able to install and troubleshoot a web-based supervisory controller and perform basic troubleshooting of an Ethernet network. As part of these tasks, the technician must be familiar with a number of basic tools used to locate, troubleshoot, diagnose, and remedy problems. These tools can be either hardware tools or software tools.

HARDWARE TOOLS

Hardware tools consist of physical devices, such as LED indicators and handheld testers. Handheld testers often feature multiple testing functions and are versatile devices. However, most troubleshooting begins with a visual check of LED indicators. This can be done easily and quickly, and it often indicates the source of the problem.

LED Indicators

The most basic, yet important, diagnostic tool is the LED indicator. While not traditionally considered hardware tools, these simple indicators can be incredibly useful during the initial steps of a troubleshooting procedure, and their value should not be overlooked. LED lights typically indicate if a device is receiving power, the device status, network connectivity, signal transmission and/or reception activity, and communication speed. LED indicators are found on NIC cards, network devices, and controllers.

NIC Card LED Indicators. The LED indicators on NIC cards are usually next to the Ethernet port. **See Figure 7-1.** One LED, known as a link indicator, lights if the card senses that it is properly connected to an active network. Another LED indicates signal activity, either sending or receiving, between the network and the card. Under normal network conditions, this LED flashes intermittently. If it is not flashing, it may indicate a hardware problem with the card, a software problem related to the card, or a network problem. Some NIC cards have a single LED to show both link and activity status.

NIC Card LED Indicators

Figure 7-1. NIC cards include LED indicators to show link and activity status.

Also, some NIC cards include additional LED indicators for communication speed, such as 100BASE-T or 1000BASE-T. Sometimes, the color of an LED is used to indicate speed.

Network Device LED Indicators. Network devices, such as hubs, switches, routers, and wireless access points, typically include many LED indicators. Indicators are next to each port for link and activity information, just as with NIC cards. The LEDs can indicate a faulty port or help isolate a network problem to a specific source. There are also LEDs that indicate power status, device status, and wireless connectivity if applicable.

Controller LED Indicators. As nodes, web-based supervisory controllers include typical LED indicators for power status, link status, activity status, and possibly connection speed. **See Figure 7-2.** Additional LED indicators are used for controller-specific features. These can vary greatly between manufacturers and models, but there are some common types.

Controller LED Indicators

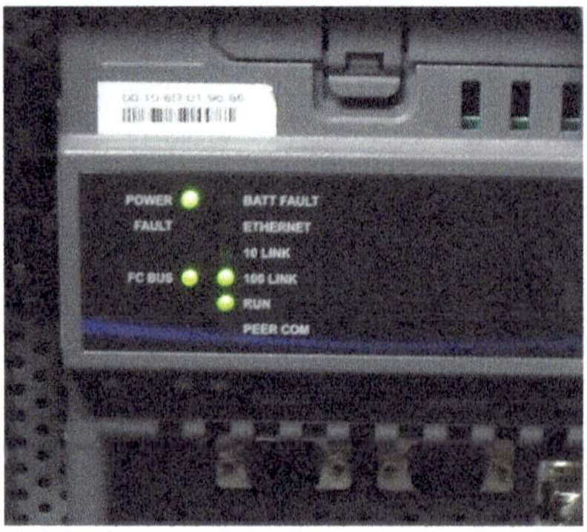

Figure 7-2. LED indicators on controllers are used for a variety of status conditions.

Indicator LEDs on network devices provide valuable information on the status of device and network functions, such as connection speed or signal strength.

A special LED may indicate an active connection to an HVAC control system server. This is likely not a direct connection via a particular port. Rather, the LED indicates that the controller is successfully exchanging information on the Ethernet network with a particular node that is identified as the server.

The web-based supervisory controller is also connected to field controllers via a communications bus that is not related to the Ethernet network. It is important to monitor these communications as well. Additional LEDs may indicate both link status and communication activity with field controllers.

Most web-based supervisory controllers have a fault LED. When lit, the fault LED indicates that there is an internal problem with the controller, though it does not indicate what the problem is. Connection through a browser or software tools must be used to determine the cause of a fault.

Many controllers include a battery, which has two functions. It keeps the device running during brief power flickers (outages lasting a few seconds) and, if an outage continues, allows an orderly shutdown of the device. Controllers with batteries typically include an LED to indicate battery status.

Handheld Testers

If the problem cannot be identified using LED indicators, HVAC controls technicians usually use handheld testers to continue troubleshooting. Handheld testers are necessary to check physical cables for continuity, damage, and proper termination. Other testers analyze network traffic, though this function can also be done with some software tools. Many handheld testers have more than one test function.

Digital Multimeters (DMMs). One of the most basic handheld diagnostic meters used is the DMM. A *digital multimeter (DMM)* is an electrical test tool that can measure multiple parameters, commonly voltage, current, and resistance. **See Figure 7-3.** While it cannot directly measure any Ethernet network parameter, it is often used to check a network device's power supply. If an initial check of LEDs indicates that there may be a power supply problem, a DMM can be used to measure the input and output voltages. Most devices use a power supply based on a step-down 120 VAC to 24 VAC transformer. Unexpected readings can help diagnose a number of power problems, such as the following:

- a circuit wired to the wrong circuit breaker
- a bad or failed power supply
- an incorrectly connected multivoltage power supply
- an incorrect power supply for a device
- inadvertently disconnected power
- an overloaded circuit from connecting too many devices
- a damaged power supply due to a lightning strike or utility power surge
- some type of intermittent power problem exists
- a blown fuse or breaker on the secondary (24 VAC) side

Digital Multimeters (DMMs)

Figure 7-3. DMMs are common electrical tools that are often used to test device power supplies during troubleshooting.

Cable Testers. A *cable tester* is a test tool that can check the integrity and proper connector pinouts of a multiconductor cable. Basic cable testers are relatively small, inexpensive units and are available for many different types of cables, including

UTP, USB, serial, and fiber-optic cables. Most cable testers consist of two units. As one unit applies a known voltage to each conductor, the other unit is used to measure the voltages at the other end to identify each conductor. **See Figure 7-4.** The tester can then detect shorts, opens, and excessive resistance on the conductors. It can also verify that the conductors are terminated in the correct order at the connectors. Some cable testers are a single unit, but these only work if there is network activity on the cable. The following problems can be detected by cable testers:

- an incorrectly crimped, field-fabricated cable
- a wrong cable run to a panel
- an incorrect cable
- a damaged cable
- a cable disconnected at the opposite end

Cable Testers

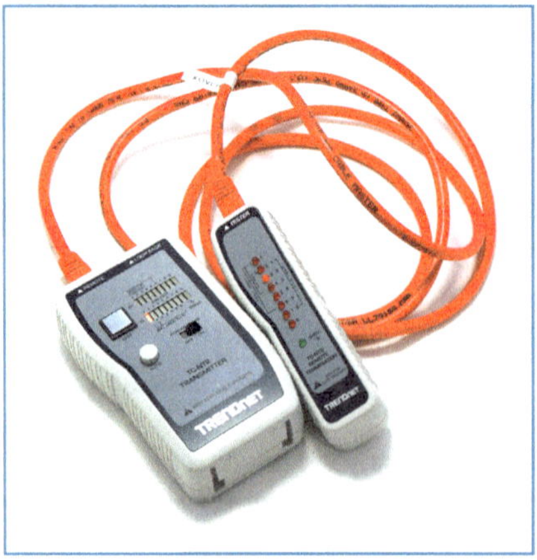

Figure 7-4. Cable testers detect common problems with wiring and connector terminations.

> Cable tracers are known by many names, including trackers, toners, and locators. They also vary in the type of connectors included, such as RJ-11 (telephone), RJ-45, or coaxial. However, nearly all cable tracers provide simple clips that can be used with any type of cable.

Cable Tracers. Network cables are often run through cableways in very large bundles. These bundles make it difficult to follow the individual cables, which may be necessary to find damaged spots or determine a cable's destination. This is especially true for the large cable trays that pass between floors.

A *cable tracer* is a test tool that can identify a particular cable out of many at anywhere along its length. **See Figure 7-5.** These testers also consist of two units. One unit connects to a cable end and generates an electrical signal on a conductor. A sniffer unit, when held near the cable, detects the slight magnetic field caused by the electrical signal and emits a corresponding audible tone. The volume of the tone typically indicates the sniffer unit's proximity to the target cable.

When a cable is being traced, cable bundles are carefully pulled apart so that the sniffer unit can be held near each one. This procedure is repeated as needed to trace a cable from one end to the other.

Network Analyzers. A *network analyzer* is a test tool that listens passively to network traffic, recording message packets and saving them for later analysis. **See Figure 7-6.** Analyzers use this information to show network data losses, detect timing issues, diagnose badly formed messages, and even show the contents of network messages.

Cable Tracers

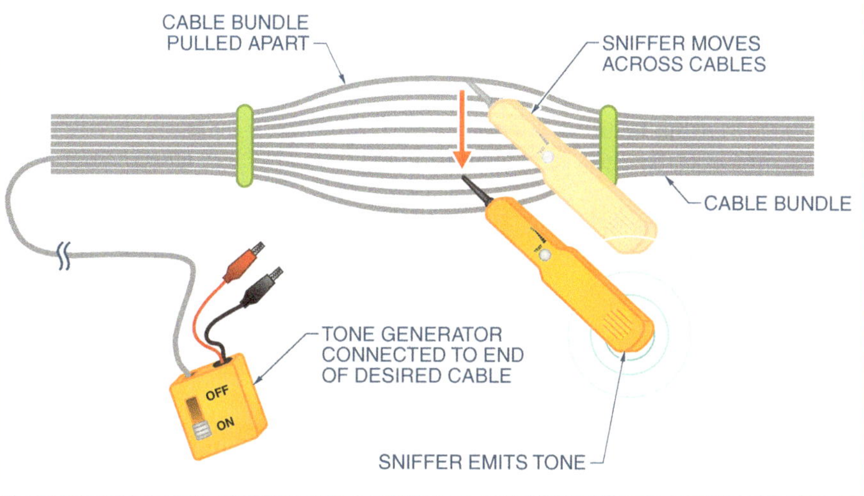

Figure 7-5. Cable tracers are used to follow or identify specific cables that are bundled with many other identical-looking cables.

Network Analyzers

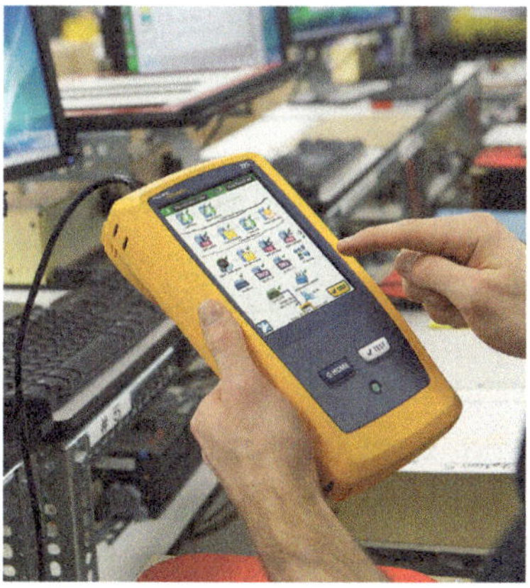

Fluke Corporation

Figure 7-6. Network analyzers are sophisticated test tools with many cable and signal testing functions. They capture and read the packets of information on a network.

A network analyzer often includes many other sophisticated testing functions, making it a handy tool for a good majority of network testing applications. Many of these tools can measure interference and crosstalk. *Crosstalk* is the induction of signals onto a conductor from an adjacent conductor. Some analyzers are also capable of detecting and locating cable breaks. While these advanced tools are often used by facility IT staff, HVAC controls technicians should consider purchasing one.

Wireless Meters. A *wireless meter* is a test tool that detects the presence and type of wireless networks and measures the corresponding signal strengths. These portable meters are used to determine problems with a facility's wireless infrastructure. Areas of insufficient wireless coverage can be detected by moving around the facility with a meter.

SOFTWARE TOOLS

In addition to hardware tools, many software diagnostic tools are available. Some simple diagnostic tools are part of a computer's OS, while sophisticated tools are available from third-party developers. HVAC controls technicians should be sure that their work laptops are configured with the proper tools and that they know how to use them.

Command Prompt Tools

A number of software tools are available through an installed OS. Many of these are accessed via the command prompt, which is a text-only interface with the OS. **See Figure 7-7.** The IT department may restrict command line access and other software tools, so technicians should check with them first.

The command prompt is opened by selecting the Windows Start button and then either selecting the Run program or typing "cmd" in the Search box. The common network configuration tools are ping, ipconfig, tracert, and netstat.

Command Prompt

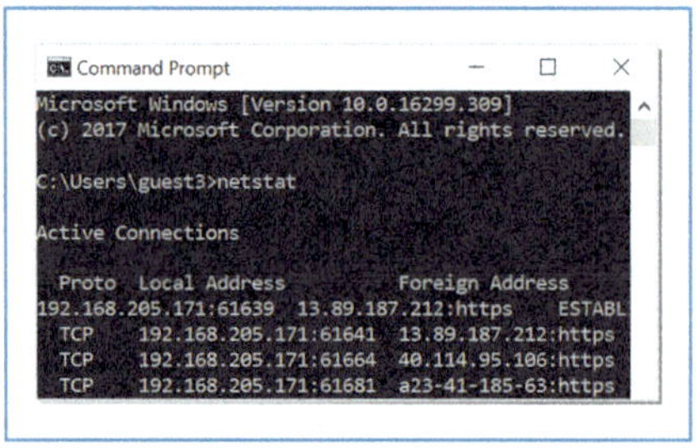

Figure 7-7. Several basic, text-only, network software tools that are included with Windows are executed from the command prompt.

The text output should be saved as part of the troubleshooting documentation. It may also be necessary to share this information with the facility IT staff in case a significant network problem requires their assistance.

Ping. The most basic and common software tool is the ping command. Network troubleshooting usually begins by "pinging" another device. The ping command tests the ability of one network device to successfully contact another. If one device can be pinged from the other, it means that communications are established between the two devices.

The ping command is executed at the command prompt by typing "`ping <ipaddress>`", using the target IP address. If communications are working between the two devices, the reply resembles the following:

```
Pinging 192.168.10.10 with
 1500 bytes of data:
Reply from 192.168.10.10:
 bytes=1500 time=68ms
 TTL=52
Reply from 192.168.10.10:
 bytes=1500 time=68ms
 TTL=52
Reply from 192.168.10.10:
 bytes=1500 time=65ms
 TTL=52
Ping statistics for
 192.168.10.10:
Packets: Sent = 3,
 Received = 3, Lost = 0
 (0% loss),
Approximate round trip
 times in milliseconds:
Minimum = 65ms, Maximum =
 70ms, Average = 67ms
```

If a ping fails, the reply is "Request Timed Out" repeated.

There are variations of the ping command, which are accessed by adding "switches" (command modifiers) to the ends of the commands. For example, "ping -t <ip address> " is a common variation that continuously pings the target until released by holding down the CTRL and C keys simultaneously. This is commonly done while working on a cable or network. When the problem is fixed, then the ping will be successful.

Ipconfig. The ipconfig (for "IP configuration") command displays the current IP address and other related addresses for the user's computer. This command is executed by typing "ipconfig" at the command prompt. Additional features of ipconfig include the ability to release and renew DHCP addresses, which can sometimes fix connection problems.

Tracert. The tracert (for "trace route") command displays the entire path from one device to another, including all routers, as well as the timing between each device. This tool may be used to help determine the topology of a poorly documented network. It can also reveal where traffic is slowing down. This command is executed by typing "tracert <ip address>" at the command prompt. The result resembles the following:

```
Tracing route to
 KMBT68168D
 [192.168.205.117]
over a maximum of 30 hops:
1    16 ms    <1 ms    <1 ms
 192.168.205.1
2    12 ms    11 ms    12 ms
 192.168.205.10
3     7 ms     8 ms     8 ms
 192.168.205.117
Trace complete.
```

Netstat. The netstat (for "network status") command displays detailed information about the devices currently communicating on the network. The information includes the communications type, local IP addresses, other address names, and statuses. In a way, netstat is like pinging every node on a LAN at once. This tool is useful for gauging the extent of any communication problems. For example, if a network includes a certain number of nodes, but only some are communicating, then reviewing which are communicating and which are not can help determine the location of the problem.

Windows Control Panel

Another method to check and/or change the IP address of a computer is through the Windows Control Panel. This function is accessed via the "Change adapter settings" section of the Network and Internet control panel. This method is especially vital if a fixed IP address is used instead of a DHCP assigned address.

Serial Terminal Capture

If the ping command is not successful, some web-based supervisory controllers have a feature that broadcasts the device's IP address on its RS-232 (serial) connection upon rebooting. **See Figure 7-8.** In order to capture this information, a computer can be connected to the serial port, and a terminal program can be used to display the data. Generally, any terminal software with a VT100 mode can be used, including HyperTerminal, which was included with older versions of Windows. The controller manufacturer can be consulted for more information on this feature and software requirements.

This data is normally read-only, so it cannot be changed via this mode. However, when it is impossible to otherwise communicate with a device with an unknown IP address, this technique can be vital for troubleshooting.

Serial Ports

Figure 7-8. Some web-based supervisory controllers include RS-232 (serial) ports, which are used for accessing configuration information when a controller is not communicating on a network.

Third-Party Network Tools

There are many software diagnostic tools available that are not part of a computer's OS. Some are available for free, though the most sophisticated ones are usually commercial products.

Comprehensive tools include many network analysis features for capturing and analyzing network traffic, much like handheld network analyzers. For example, Wireshark® is a good network analyzer utility. Data from these network tools is often sent to controller manufacturers for analysis as part of technical support.

Vendor Tools

Manufacturers of web-based supervisory controllers also make tools that can be used to troubleshoot and service their specific controllers. These tools are used to find the controllers on the network and identify their IP addresses and software versions. Often, the web browser can be launched from the utility to gain access to the controller as well. It should be noted that these tools may need to be updated as the vendors update their controllers. HVAC controls technicians may want to load these tools on an on-site facility computer as well as on their own service laptop.

Review Questions

1. How can LED indicators be useful for initial troubleshooting?
2. What are some types of LED indicators commonly found on web-based supervisory controllers?
3. What part of a network device can be tested with a DMM?
4. What are the common types of problems detected by cable testers?

5. What tool includes multiple network testing functions?
6. What is the command prompt and how is it used for network troubleshooting?
7. How can the tracert command be useful for network troubleshooting?
8. What are RS-232 (serial) connections on web-based supervisory controllers sometimes used for?
9. What are some other sources of network software tools?

8 Troubleshooting Scenarios

Accurate and effective troubleshooting skills are an HVAC technician's greatest assets. These skills distinguish the outstanding technicians from the average ones and often lead to more-frequent promotions and higher pay.

TROUBLESHOOTING

There are numerous ways a network could be designed, making it nearly impossible to provide step-by-step troubleshooting instructions for every possible problem. In addition to having a good understanding of networking principles, troubleshooters must be adaptable to a wide variety of situations. This makes the critical skill of troubleshooting difficult to master. Fortunately, there are general guidelines that can help technicians resolve many of the most common issues likely to arise in the field.

Troubleshooting Skills

Generally, the process of troubleshooting continually narrows the location and possible causes of a problem until the root cause is identified. First, the symptoms of a problem are identified. Then, troubleshooters rely on knowledge, experience, and other resources to guess at possible causes and their locations in the system.

Testing is then used to eliminate the parts of the system without the problem. It is helpful to think of a system, any system, as a group of discrete sections. The points between sections are where troubleshooting tests or inspections can be performed to determine if a section is functioning properly. Technicians can then apply the process of elimination to determine where the cause of a problem is likely to be.

For example, a computer network that includes an HVAC control system may have the following sections: between the client computer and the wall jack, between the wall jack and the router or switch, between the router or switch and the server, between the router or switch and the web-based supervisory controller, and so on. **See Figure 8-1.** Additionally, the software used with any computer, server, smartphone, web-based supervisory controller, or similar device, as well as the configurations of those devices, could each be considered a section.

If there is a communication problem with a client computer, for example, but tests reveal a normal signal from the wall jack, then the problem likely involves either the computer's software or the physical infrastructure between the computer and the wall jack, such as the NIC card or the patch cable. The rest of the network is likely working properly, and no further troubleshooting is required beyond the wall jack. Knowing this information, the technician can then conduct additional tests on the computer.

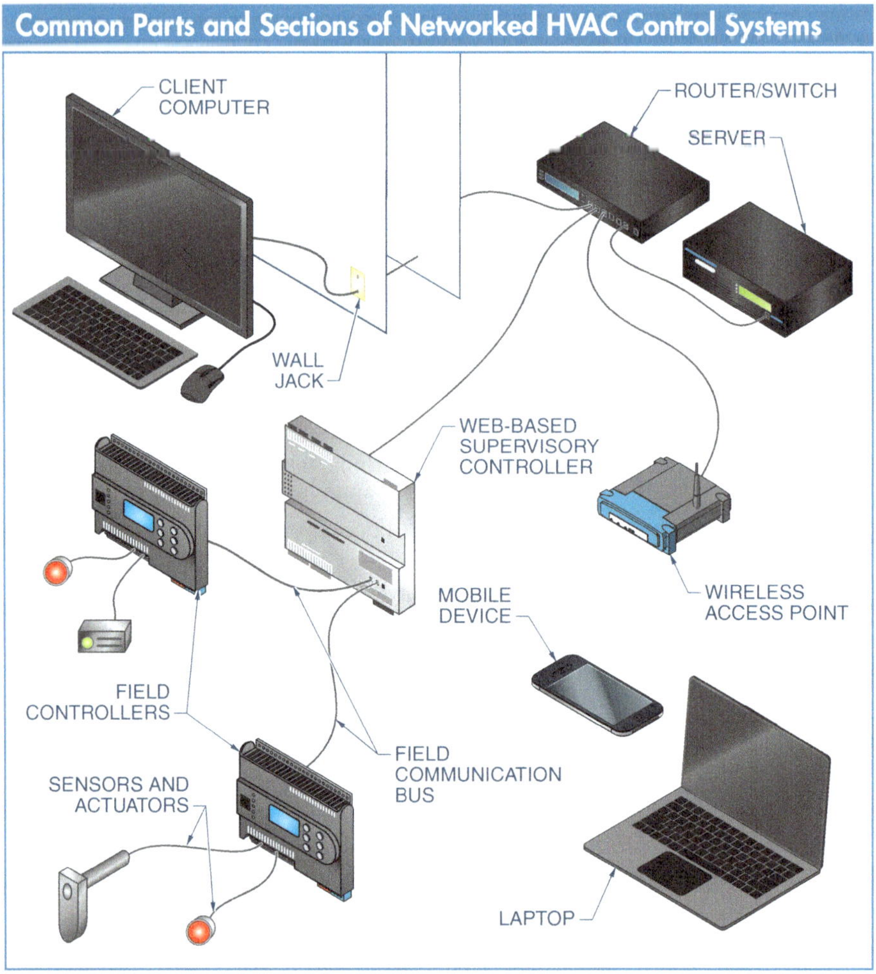

Figure 8-1. When troubleshooting is performed on a networked HVAC control system, the possible locations of a problem are gradually eliminated down to a single piece of equipment or section.

Technicians should investigate and rule out the most likely and most easily detected (or eliminated) causes first. Troubleshooting is often required under time constraints, so it is important for technicians to prioritize their investigations. While random and obscure failures can happen with any equipment, they are rare. No time should be wasted tracking down unlikely causes until more likely causes are eliminated first.

Troubleshooting Resources

To accurately and quickly troubleshoot a problem with a networked HVAC control system, the technician should have the following resources available:

- access to controller, server, and network device locations
- a list of assigned IP addresses and network names for all devices
- passwords for client computers, servers, and wireless networks, as needed
- diagrams of network infrastructure and locations of cable runs
- hardware and software diagnostic tools, as needed
- instruction sheets and manuals for network devices
- backup copies of software

With any HVAC system, it is imperative to get as much pertinent information as possible. This can often lead to a quick, easy diagnosis and fix. An experienced technician asks many questions to help find problems quickly. Examples of good questions to ask include the following:

- When did the problem begin?
- Has it happened before?
- When was the last time the system worked properly?
- Have any changes been made to the system recently?
- Have any parts been replaced or added?
- Has anyone worked on the system recently?

COMMON SCENARIOS

While network problems have many possible causes, most problems are due to a relatively small number of common ones. **See Figure 8-2.** Most network problems are predictable and repetitive and can be resolved by skilled and knowledgeable HVAC technicians. The following are actual troubleshooting scenarios from real jobs as performed in the field by a senior HVAC controls technician.

Incorrect Client Computer Setup

A customer complains that a client computer cannot access the web-based HVAC control system. While interviewing the user, the technician learns that the computer was recently installed to replace an older one. It is then revealed that the computer was never set up to access the control system. The technician confirms that the computer cannot access, or even ping, the web-based supervisory controller.

The technician uses the Windows Control Panel to access the network settings and finds that, instead of fixed IP, the computer is set up to use DHCP. The technician then consults the system documentation for the correct IP address, subnet mask, and router address. After this information is entered and saved, the computer can now ping the web-based supervisory controller. The technician uses the browser to access the controller and confirms that access is restored.

Advice: When installing or servicing a web-based HVAC control system, technicians should inform customers that, if they want to access the system from an additional client computer, the computer must be set up properly first.

Bad Patch Cable at Client Computer

A customer complains that the web-based supervisory controller cannot be accessed from a client computer. While asking about any recent changes, the technician learns that the computer worked previously but has not since it was recently moved to another location. The technician confirms that the computer cannot access or ping the controller.

Common Troubleshooting Scenarios		
Initial Problem	**Other Notes**	**Possible Causes**
One client computer cannot access the network at all; other computers are working properly.	Computer was recently installed.	Incorrect client computer setup
		Incorrect cable type
	Computer was recently moved (access worked before it was moved).	Bad patch cable at client computer
		Incorrect network port
	Work was recently performed in telecom closet.	Bad cable in telecom closet
		Disconnected cable at switch/router
One client computer cannot access the local HVAC control system, but can still access the Internet; other computers are working properly.	Computer is using a wireless network connection.	Incorrect wireless network
	Pinging is successful.	Blocked browser pop-ups
	A different browser is being used.	Incompatible browser
	Java software was updated.	Incorrect Java software version
Controller cannot be accessed over the network by any computer.	Pinging is unsuccessful; controller was recently configured.	Incorrect controller IP address
	Pinging is successful.	Duplicated IP address
	Router or switch was recently replaced.	Incorrect connection speed
	Software was recently updated on all computers.	Incompatible software updates
Server cannot be accessed over the network by any computer.	Server was recently installed.	Incomplete server setup
	Server's firewall is enabled.	Access blocked by firewall
User cannot log into computer.	Operating system password is invalid.	Incorrect computer password
Laptop or mobile device cannot access the HVAC control system remotely.	VPN is used to go through firewall.	Incorrect VPN software version
	Device is not connected to wireless network.	Incorrect wireless password

Figure 8-2. Most network problems involving HVAC control systems are due to a relatively small number of causes.

The technician visually checks the Ethernet patch cable between the wall jack and the computer's NIC card. The cable is properly connected to the correct port, but the cable appears to be damaged. A cable tester is attached to the end of the cable, and the tester indicates an open circuit on one of the conductors. **See Figure 8-3.** While there are many possible causes, the visible damage on the cable indicates that the cable itself most likely needs to be replaced. Also, cable damage is not uncommon when moving computers or furniture. The technician replaces the cable with a new one, and the computer is then able to access the system.

Advice: Technicians should stock some new patch cables in their service vans. These are inexpensive items that can be used to quickly and easily solve commonly encountered problems, or at least verify that a bad cable is not the cause. Cables are available at almost any retailer with office or computer equipment. Long cables, such as 20 ft cables, are recommended in case extra length is needed to reach the nearest wall jack. Long cables can always be coiled and secured with wire ties until they can be replaced with appropriate-length cables as a permanent fix.

Testing Patch Cable

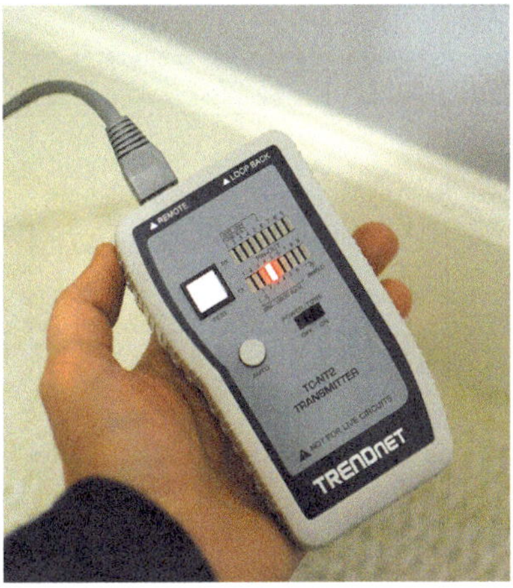

Figure 8-3. A cable tester can be used to test the wiring of a patch cable.

Bad Cable in Telecom Closet

A customer complains that a client computer cannot communicate with the HVAC control system. The problem started the previous day, and all the other computers are working normally. The technician confirms that the client computer cannot access the HVAC control system. The computer also cannot ping the web-based supervisory controller.

The technician checks the cable connected to the computer with a cable tester, but no signal is present. There is also no signal present at the wall jack. The technician traces the cable back to the telecom closet, where the end is connected to the switch. However, no activity or link LEDs are lit on the port the cable is connected to. Other switch ports in use have lit LEDs, suggesting that the problem is likely not with the switch, so attention is focused on the cable. The customer then mentions that the IT staff was working in the telecom closet the previous day. The technician now considers a damaged cable to be the likely cause of the problem.

The technician tries to visually inspect the length of the cable, but it is joined with many others in a large bundle, making the cable almost impossible to trace. The technician uses a cable tracer to identify the correct cable within the bundle. The tone generator unit is connected to one end of the cable, and the sniffer unit is passed over the bundle. **See Figure 8-4.** The correct cable is identified and separated from the bundle. When examined, it is found to be damaged, likely by the recent work performed by the IT staff. When the cable is replaced, the switch port LEDs light, and the client computer can then access the controller.

Advice: Work done by a facility's IT staff can affect a network-based HVAC control system, so it is important for technicians to maintain a working relationship with the staff. Troubleshooting in the telecom closet and inspecting cable bundles requires the cooperation and assistance of the IT staff. HVAC controls technicians should keep the IT staff's contact information available in case such a situation arises.

Tracing Cable

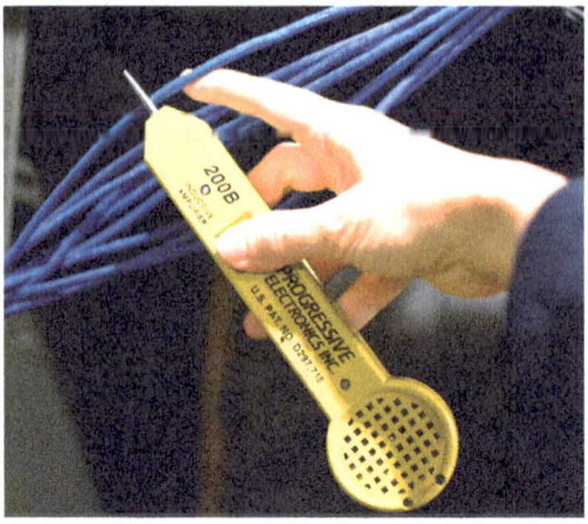

Figure 8-4. A cable tracer can be used to identify a particular cable from a bundle of similar-looking cables.

Incorrect Cable Type

A customer complains that a recently installed client computer cannot access the HVAC control system. The computer's network settings have been configured, but network access was never confirmed. The HVAC controls technician confirms that the computer cannot access the control system or ping the web-based supervisory controller.

The technician visually inspects the cable at the computer. A close look at the printing on the cable indicates that it is a crossover cable and not a patch cable. **See Figure 8-5.** While some computers can automatically detect and reconfigure their signaling for the different pin-outs of a crossover cable, many cannot. For these computers, an incorrect cable type would prevent communication. The technician realizes that a crossover cable was mistakenly used by an installer, and the connection was never verified. After the cable is replaced with a proper patch cable, the computer is then able to successfully communicate on the network.

Advice: UTP cables are sometimes marked or labeled. Some manufacturers designate certain cable insulation colors for the two types, but this is not a standard practice in the industry. Also, field-fabricated cables may not be color-coded, or they may be terminated incorrectly. Therefore, technicians should carefully check any labeling in order to choose the right cable. When in doubt, technicians should verify the order in which conductors are terminated for both crossover and patch cables.

Cable Markings

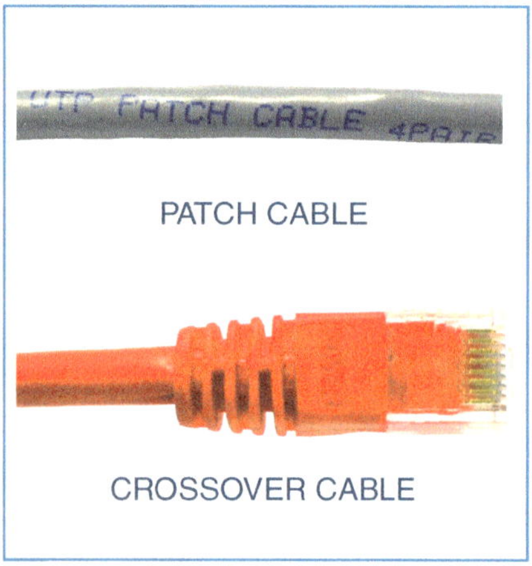

Figure 8-5. Some crossover cables are marked on the insulation.

Disconnected Cable at Switch/Router

The customer complains that a client computer can no longer access the HVAC control system. The problem started the previous day. In addition, a network printer is no longer working. The computer and printer are confirmed to have no active connection to the rest of the network.

The HVAC controls technician contacts the facility's IT staff to request access to the telecom closet and asks if any work has been performed recently. The staff confirms that a router was worked on the previous day. It is immediately clear to the technician, upon checking the router in the telecom closet, that the cables had been disconnected from the router and not reconnected. The cables for both the client computer and the printer are plugged into the correct places, and the indicator lights flash as expected. **See Figure 8-6.** The client computer is then able to access the system properly.

Advice: Problems are frequently caused by recent changes to systems, so technicians should always ask about any such changes when troubleshooting. However, work performed by the IT staff is not typically known to users, so technicians may occasionally need to speak directly with IT staff to solve problems.

Checking Router Connections

Figure 8-6. Cable connections and LED indicators can be checked at router ports.

Incorrect Controller IP Address

An HVAC controls technician installs a new web-based supervisory controller and configures it with the fixed IP address provided by the facility's IT staff. After installation and startup, the controller cannot be accessed from a client computer in the maintenance shop. The technician tries to ping the controller and is unsuccessful, receiving the error message "destination host unreachable."

There have been no recent changes to the system. The client computer can still access other nodes, and the LEDs at the switch, router, and controller ports are lit. This indicates that the physical network infrastructure is operating properly. The technician suspects that the problem is with the controller's network configuration. The technician compares the client computer's IP address configuration with the one provided for the controller and finds that they are on different subnetworks. The technician checks with the IT staff and discovers that the wrong IP address was inadvertently provided.

Since the controller cannot be reached through any client computer, the IP address cannot be changed using them. Instead, the technician disconnects the controller from the building network and sets up a small, temporary network with just the controller, a laptop, and a small switch. **See Figure 8-7.** The IP address is reconfigured properly. When reconnected to the building network, the client computer is then able to access the web-based controller.

Advice: It is very easy for an IP address to be incorrectly heard, read, documented, or typed. An error in just one digit can prevent a node from communicating on a network. When receiving IP addresses from IT staff, technicians should repeat the numbers back to the staff in order to confirm the digits. Also, if IP addresses are written down, technicians should always check with the IT staff to confirm addresses before making any configuration changes. For example, all network equipment should be labeled with IP addresses and other identifications, but this information could be outdated and incorrect.

Incomplete Server Setup

An HVAC controls technician installs a number of web-based supervisory controllers, along with a server to store data. The server is supplied by the IT department and is located in the telecom closet in a rack mount server cabinet. Following installation, the technician tests communication with the server and discovers that the client computer cannot access the server.

The technician confirms access to all of the individual controllers from the client computer. In the telecom closet, the connections between the server and the switch are checked. All LED indicators are lit as expected. This suggests that the problem is with the server configuration. The technician then logs on to the server to check the network settings and controls software. It is then discovered that the log-on credentials used by the technician did not allow full administrative rights, which prevented the software from being loaded properly. The technician contacts the IT staff for a full administrative password, which is then used to properly load the software. The server is then accessible on the network and operating.

Setting up a control system server involves configuring network settings and loading application-specific software.

Reconfiguring Controller IP Address

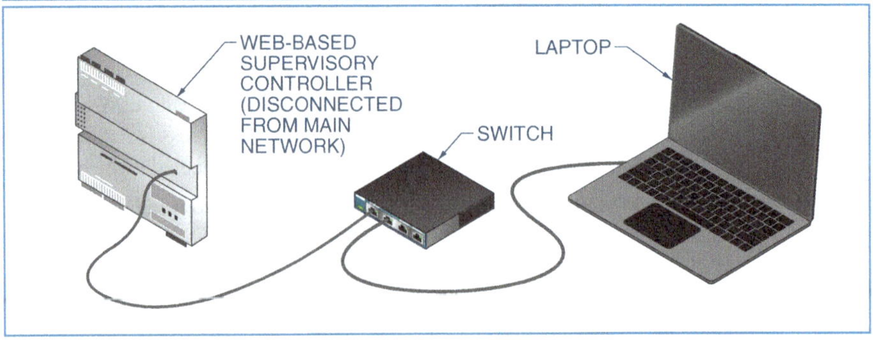

Figure 8-7. Technicians should always have a small network switch available for installations or troubleshooting. The switch can be used to make a small, temporary network in order to configure a web-based supervisory controller directly.

Advice: One of the primary roles of the IT staff is to provide system security, which is critical to maintaining system integrity. This means that they may be reluctant to grant full administrative rights to their equipment, as it may compromise security. However, this prevents software from being installed and loaded properly, hampering HVAC operations. HVAC controls technicians need to work with the IT staff on possible solutions. For example, one option is for the staff to provide temporary administrative rights, which are then revoked once the HVAC system is installed and operating.

Incorrect Wireless Network

A customer complains that a laptop can no longer access the HVAC control system via the building's wireless network, though it was operating properly the previous day. The technician confirms that the laptop cannot access the local HVAC control system, though it still has access to the Internet. The wired computers at the facility can still access the system.

Given the circumstances, the problem is likely related to the wireless network. However, the wireless icon in the system tray indicates a normal connection and sufficient signal strength. The technician then considers that the laptop can access the Internet only and the wired computers can access both the Internet and the HVAC system. Checking the laptop further, the technician discovers that it is connected to the public wireless network instead of the private network, which includes the HVAC system. **See Figure 8-8.** The technician disconnects the laptop from the public wireless network and connects it to the internal private wireless network, and the laptop is then able to access the HVAC system properly.

Advice: It is common for facilities to have more than one wireless network, and the networks may have different levels of access to local devices. Portable devices, such as laptops, may lose a connection to one wireless network and automatically reconnect to another, depending on signal coverage. When troubleshooting wireless network problems, technicians should always inquire if there are multiple networks and verify which network a computer is actually connected to.

Selecting Wireless Networks

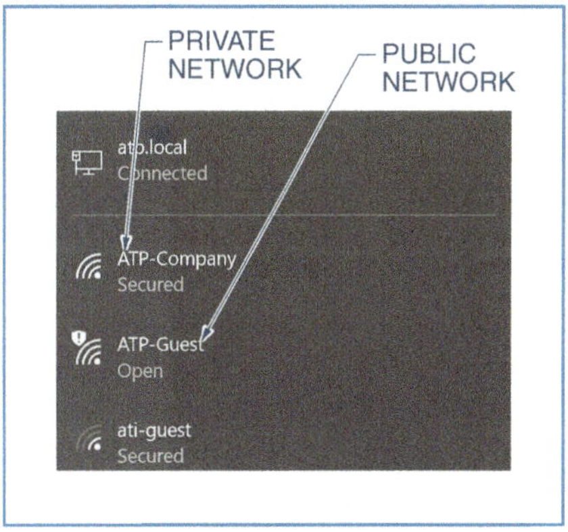

Figure 8-8. The wireless network panel shows which network a computer is actually connected to, along with any other available networks in range.

Access Blocked by Firewall

An HVAC controls technician installs an HVAC control system with a server but discovers that the system cannot be accessed from any client computer. The server computer was provided by the facility's IT staff.

The technician consults a troubleshooting checklist provided by the controls vendor. According to the checklist, one of the first items to check is the firewall. The technician notices that the Windows firewall icon in the system tray indicates that the firewall is enabled. The technician accesses the firewall settings from the Control Panel and learns that the firewall is not set up to allow access to the HVAC control system server software. The technician works with IT staff to change the firewall settings to allow client computers to access the software. After these changes, the server is rebooted, and access is then allowed.

Advice: While firewalls are critical for system security, they are very conservative and block all communications that are not specifically approved. Changing the settings to allow control system communications requires cooperation with the IT staff.

Incorrect VPN Software Version

A customer has been using a smartphone to access an HVAC control system. A few days ago, this stopped working. The virtual private network (VPN) that is used to ensure a secure connection no longer allows the user to log on.

All on-site computers are able to access the HVAC control system, so the problem seems to involve the VPN. The HVAC controls technician contacts the facility's IT staff about the VPN and learns that the VPN software has been updated recently. The customer's smartphone did not receive the update, and the conflicting software versions are preventing access. The customer's smartphone VPN is updated, and then access to the HVAC system is restored.

Advice: A facility's IT department will often use a VPN in order to provide a secure means of connecting to the LAN remotely. The IT staff must set up the software, provide log-in credentials, and test the connection to be sure it is working properly. VPNs are a common source of potential problems, however, as any changes to settings or software versions can easily disable access.

A virtual private network (VPN) is a way for remote users to connect to and work on a LAN as if they were physically connected to the LAN.

Blocked Browser Pop-Ups

A customer tries to set up a new client computer to access an HVAC control system but is unsuccessful. All required software has been installed, but the browser displays no data from the web-based supervisory controller.

The HVAC controls technician successfully pings the server and demonstrates that the network connection is active and functioning. The technician checks the computer's network settings, and they appear to be correct. The technician tries to access the control system via the browser again and notices that the window seems to flash briefly. The technician suspects that browser settings are the problem. It is discovered that the browser's pop-up blocker is turned on and has blocked a number of pop-ups from the control system's website. **See Figure 8-9.** The technician disables the pop-up blocker and restarts the browser. The client computer now functions properly and is able to access the system.

Accessing Pop-Up Blocker Settings

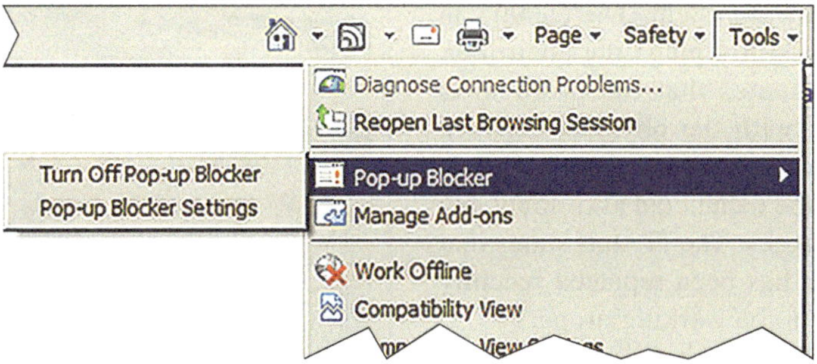

Figure 8-9. Browser pop-up blockers may need to be set to allow pop-ups from the control system's controllers and server.

Advice: The browser used on a computer, tablet, or smartphone must be supported by the manufacturer of the web-based control system. In addition, settings related to Java, pop-up blockers, and security must be configured properly. If the client computer is used for more than just accessing the control system, the browser should be set to block pop-ups from all websites except the address of the control system.

Duplicated IP Address

A customer complains that a client computer cannot access the web-based supervisory controller. The technician confirms the problem but is able to successfully ping the controller. The physical network infrastructure seems to be working properly.

First, the technician tests the controller by disconnecting it from the building network and attempting to log on to the controller with a laptop and switch. The local log-in works properly. The technician then suspects that the problem is with the network settings. Connecting the laptop, but not the controller, to the building network, the technician pings the IP address successfully. Since the controller is disconnected from the network, the only possible explanation is that the controller's IP address is being used by another node on the network. The technician contacts the IT department and discovers that a new printer has been inadvertently assigned the same IP address. The printer's IP address is changed, and the controller is reconnected to the network. The client computer is then able to access the control system.

Advice: The IT staff maintains lists of IP addresses in use, but web-based supervisory controllers may not be included, which can lead to IP addresses being duplicated. HVAC controls technicians and IT staff should share device and IP address lists to avoid this problem.

Incorrect Connection Speed

A customer complains that a client computer can no longer access the HVAC control system. The technician first successfully pings the controller, which indicates that the problem is likely not with the physical network infrastructure. Since this is a new problem, the technician asks about any recent changes. The IT staff states that the switch has been replaced recently but seems to be working properly.

The technician then decides to use a network analyzer to inspect the network traffic. The analyzer shows that the connection speed is 1 Gbps. **See Figure 8-10.** However, the web-based supervisory controller has a maximum data speed of 100 Mbps. While the connection speeds of switches and controllers should automatically synchronize, the switch was not doing so in this case. The technician then reboots the controller, which forces it to reestablish communication with the switch. The connection speed of the switch then successfully synchronizes with the connection speed of the controller, and the computer is able to communicate with the HVAC control system.

Advice: When a networked control system is being installed or serviced, it is best to start up or reboot a controller after all the routers and switches are operating. This synchronizes the connection speeds of the network devices.

Many network devices, such as routers and switches, support multiple connection speeds simultaneously. The speed of each port synchronizes with the speed of its connected node to achieve the best speed possible.

Checking Connection Speeds

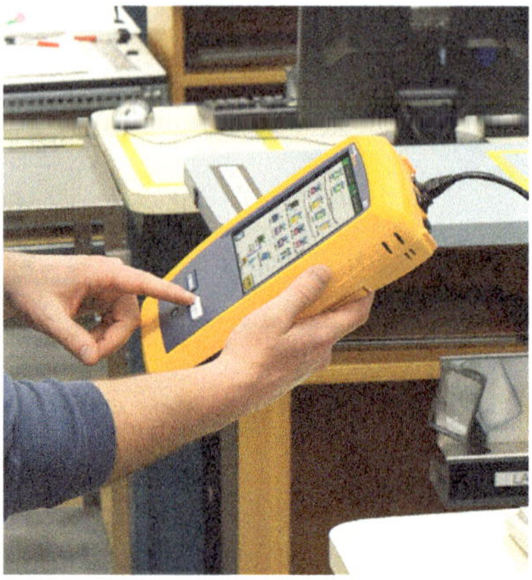

Fluke Corporation

Figure 8-10. A network analyzer can be used to check the connection speed of a network section.

Incorrect Wireless Password

A customer cannot access the HVAC control system from a tablet computer, though the tablet can access the Internet. Other wired and wireless computers are able to access the system normally, suggesting that the problem involves the tablet.

The technician reviews the network settings on the tablet and sees that it is connected to the public wireless network instead of the private wireless network, which includes the HVAC control system. The technician attempts to connect to the private network but is prompted for a password. The password provided by the user does not work. The technician consults the IT staff and learns that the wireless network password was changed but the customer was not informed. When the correct password is used, the tablet is able to access the control system.

Advice: Many wireless networks, especially private networks, are password-protected to prevent unauthorized access. Many IT departments change the passwords periodically for security reasons.

Incorrect Network Port

A client computer is moved to a new location in a building. After it is moved, however, it can no longer access the HVAC control system. The computer cannot ping any other device on the network, which suggests a physical network problem involving the computer's new location. The technician inspects the connected Ethernet port on the computer and observes that the LED indicators are not lit. A second Ethernet port is unused.

The technician uses a cable tester at the network wall jack and confirms that the cable back to the switch is wired correctly and not damaged. The technician then reviews the network settings for each NIC card in the Windows Control Panel. One of the NIC cards is active, and the other has been disabled. The technician realizes that, after the computer was moved, the cable was inadvertently plugged into the wrong card. **See Figure 8-11.** The technician plugs the cable into the other card's Ethernet port, and the computer is then able to access the control system.

Advice: Most computers have a single NIC card and Ethernet port. However, it is easy to add a second port, which may be required for some applications. In these cases, if the Ethernet cable is plugged into the wrong card, the computer will be unable to communicate with the system. If a computer has more than one physical port, technicians checking network settings should be sure that the correct port's settings are reviewed.

Inspecting Network Ports

Figure 8-11. Computers may have more than one network port.

Incompatible Software Updates

A customer cannot access the web-based HVAC control system on any of the facility's computers. This problem started the previous day.

First, the HVAC controls technician asks the IT staff about any recent changes and learns that every computer in the facility recently had software updates. The technician reviews the controls software documentation, consults the controls manufacturer, and discovers that some of the software updates are incompatible with the control system. The updates are uninstalled on some client computers. When these computers are restarted, they are once again able to access the control system.

Advice: It is recommended to have IT staff disable automatic updates to any control system client computer. Updates can always be manually installed later if they are determined to be compatible with the control system.

Incompatible Browser

A customer cannot access the HVAC control system from a client computer. When the customer attempts to do so, the browser is blank and does not display any information. The computer can successfully ping the web-based supervisory controller, and other computers can access the web interface with no issues. The problem seems to involve only that particular computer's software or settings.

The technician notices that the customer is using a different browser than the one used on the others and, after asking about it, learns that the customer prefers that browser. The technician reviews the system's technical literature and sees that only one browser is supported for the control system interface. When the standard browser is launched, the system can be accessed normally. The customer is instructed to use the standard browser when trying to access the HVAC system.

Advice: Many web-based supervisory controllers support multiple browsers, but some only support a particular one. Also, different versions of the same browser may or may not be supported. It is important for technicians to be sure a customer is aware of which browsers are supported by the HVAC control system.

Incorrect Computer Password

A customer complains that they cannot log on to a client computer. Upon arrival, the HVAC controls technician learns that the problem does not involve the HVAC control system but that the user cannot log on to the computer at all. The supplied password is invalid.

The technician contacts a member of the IT staff, who explains that the passwords are changed every 90 days for security reasons. An updated password is supplied, and the computer is then able to operate normally. The technician also confirms that the control system can be accessed.

Advice: HVAC controls technicians are sometimes called to troubleshoot computer problems unrelated to the control system, especially if the technician has been particularly helpful to the user in the past. Often, the reason is simply because the user is unaware of who to call. Technicians usually try to help in these situations but may want to discuss with the user how to determine whether the facility's IT staff or the HVAC technician should be called for future issues.

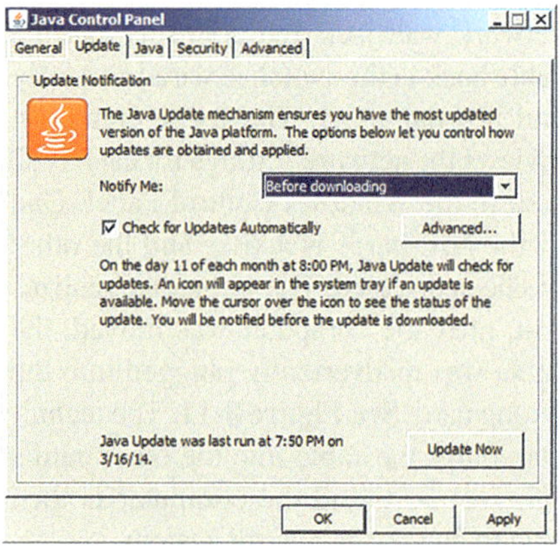

Automatic Java software updates can be enabled and disabled from the Windows Control Panel.

Incorrect Java Software Version

A customer complains that a client computer can no longer access the HVAC control system. This computer functioned properly until recently. An HVAC technician confirms the problem but is able to ping the controller from the computer. The technician is also able to log on properly using the service laptop, suggesting that this is a configuration problem with the computer.

The technician checks the network settings, and they seem to be correct. The technician also checks the software installed and notes the versions. While reviewing the controller manufacturer's software requirements, the technician discovers a discrepancy in the versions of Java. The technician realizes that the customer must have inadvertently updated to a newer version of Java that was not compatible with the HVAC control system. The technician installs the correct version from a flash drive. The client computer is then able to access the control system.

Advice: Many software applications for web-based supervisory controllers support multiple versions of Java. However, incompatible versions of Java are still common problems, so HVAC controls technicians should have installers for various versions readily available (such as on a flash drive).

Review Questions

1. What is the general process of troubleshooting?
2. What possible causes should be investigated first?
3. What are some good questions to ask in order to collect information quickly about a problem?
4. Why is it important for technicians to have replacement patch cables available?
5. How can crossover cables cause communications problems?
6. Why is it important to confirm and document an IP address provided by the IT staff?
7. How can a computer end up on the wrong wireless network, preventing communication with the HVAC control network?
8. What are the symptoms of an IP address being used for multiple nodes?
9. What action is recommended in order for routers and switches to synchronize connection speeds with web-based supervisory controllers?
10. Why should automatic software updates be disabled for client computers?
11. Why are HVAC controls technicians sometimes involved in troubleshooting computer problems unrelated to the HVAC controls system?

Appendix

Metric Prefixes ____ 89
Twisted-Pair Cable Abbreviations ____ 90
Twisted-Pair Cable Termination Color Codes ____ 90
OSI Model Analogy ____ 91
Network Device Symbols ____ 92
Numbering Systems ____ 93
Numbering System Designations ____ 94

Metric Prefixes			
Multiples and Submultiples	**Prefixes**	**Symbols**	**Meaning**
1,000,000,000,000 = 10^{12}	tera	T	trillion
1,000,000,000 = 10^{9}	giga	G	billion
1,000,000 = 10^{6}	mega	M	million
1000 = 10^{3}	kilo	k	thousand
100 = 10^{2}	hecto	h	hundred
10 = 10^{1}	deka	da	ten
1 = 10^{0}			
0.1 = 10^{-1}	deci	d	tenth
0.01 = 10^{-2}	centi	c	hundredth
0.001 = 10^{-3}	milli	m	thousandth
0.000001 = 10^{-6}	micro	µ	millionth
0.000000001 = 10^{-9}	nano	n	billionth
0.000000000001 = 10^{-12}	pico	p	trillionth

Twisted-Pair Cable Abbreviations			
Industry Acronyms	**ISO/IEC 11801 Name**	**Cable Shielding**	**Pair Shielding**
UTP	U/UTP	none	none
STP, ScTP	–	none	foil
FTP, STP, ScTp	–	foil	none
STP, ScTP	S/UTP	braiding	none
SFTP, S-FTP, STP	SF/UTP	braiding, foil	none
FFTP	F/FTP	foil	foil
SSTP, SFTP, STP	S/FTP	braiding	foil
SSTP, SFTP	SF/FTP	braiding,foil	foil

The code before the slash designates the shielding for the cable itself, while the code after the slash determines the shielding for the individual pairs:

U = Unshielded
F = Foil shielded
S = Braided shielding (outer layer only)
TP = Twisted Pair

Twisted-Pair Cable Termination Color Codes			
TIA 568A		**TIA568B**	
Pin #	Color	Pin#	Wire Color
1	White/Green	1	White/Orange
2	Green	2	Orange
3	White/Orange	3	White/Green
4	Blue	4	Blue
5	White/Blue	5	White/Blue
6	Orange	6	Green
7	White/Brown	7	White/Brown
8	Brown	8	Brown

OSI Model Analogy

The OSI Model divides the process of each network transmission into a list of successive tasks, assigning a layer of responsibility for each one. This simplifies the implementation of a particular function and makes it somewhat independent of the other layers. However, the distinction between the responsibilities of the individual layers can be confusing. To understand each layer's function, it is often useful to apply the concepts in the OSI Model to a type of communication that is easier to relate to. For example, consider the following situation in which important messages are exchanged between company offices.

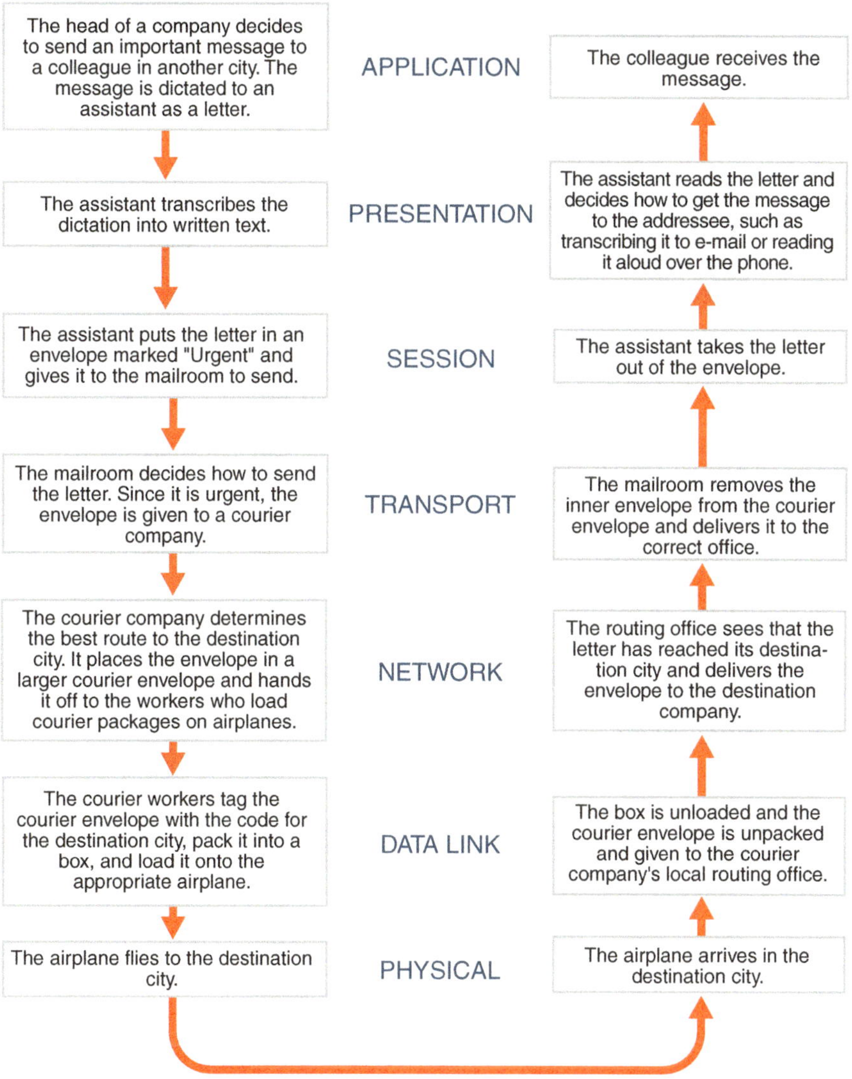

Network Device Symbols

Network devices are represented in network diagrams by symbols that indicate their respective functions. The telecommunications industry has not standardized these symbols, but many are used so commonly that they have become de facto standards. Variations are commonly used to denote devices with special features or a combination of network functions.

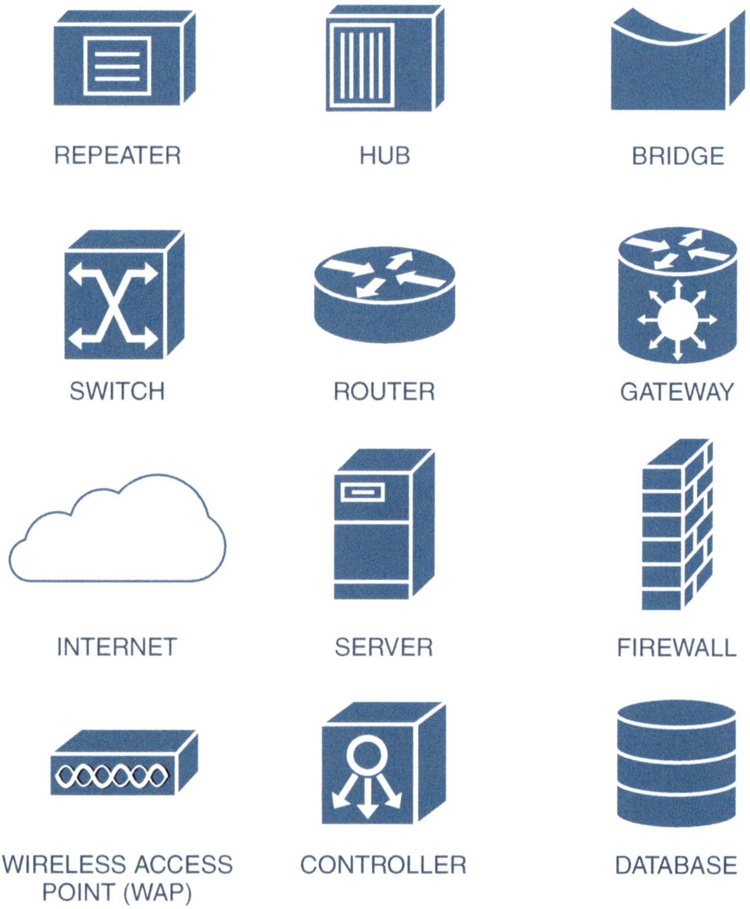

Numbering Systems

A numbering system is a unique way to represent numerical quantities with a certain range of symbols. The same number can be represented in different ways by different numbering systems. Every numbering system is defined by a base that determines the total number of symbols used to represent quantities.

Decimal	Binary	Binary Coded Decimal	Octal	Hexadecimal
1	1	0001	1	1
2	10	0010	2	2
3	11	0011	3	3
4	100	0100	4	4
5	101	0101	5	5
6	110	0110	6	6
7	111	0111	7	7
8	1000	1000	10	8
9	1001	1001	11	9
10	1010	0001 0000	12	A
11	1011	0001 0001	13	B
12	1100	0001 0010	14	C
13	1101	0001 0011	15	D
14	1110	0001 0100	16	E
15	1111	0001 0101	17	F
16	10000	0001 0110	20	10
17	10001	0001 0111	21	11
18	10010	0001 1000	22	12
19	10011	0001 1001	23	13
20	10100	0010 0000	24	14
30	11110	0011 0000	36	1E
45	101101	0100 0101	55	2D
68	1000100	0110 1000	104	44
99	1100011	1001 1001	143	63
671	1010011111	0110 0111 0001	1237	29F
1876	11101010100	0001 1000 0111 0110	3524	754

Numbering System Designations

When there is the possibility of confusion between different numbering systems, subscripts or symbols are added to distinguish the base. These are not standardized, so there are several different designations that are commonly used. The most common may depend on the industry or field using the designations.

Decimal is often assumed to be the default numbering system, unless the context specifically explains otherwise. When it is necessary to designate decimal numbers, such as when they are used in combination with numbers in other systems, decimal numbers often use "dec" as a prefix or suffix. They may instead use a subscript "10" after the number.

Binary numbers are commonly indicated with "bin" or "b" as a prefix or suffix, subscript "2" after the number, or the symbol "%." The "%" has become a common convention because the symbol resembles the numeral one with two zeroes.

Octal numbers are commonly indicated with an "oct" subscript or an "o" prefix, such as "47oct" or "o47." However, the "o" can be confused with 0 (zero).

Hexadecimal numbers are commonly written with a "0x" or "h" prefix, "hex" subscript, or "#" or "h" suffix. The "0x" prefix is the most common.

Decimal	Binary	Octal	Hexadecimal
37	0b100101	0o47	0x56A
37_{10}	10010_{12}	47_{8}	$56A_{16}$
37dec	bin 100101	47oct	56Ahex
	100101b	o47	h56A
	%100101		56A#
			56Ah

Glossary

A

address: A node's unique identification code.

alarm: A notification that the value of a system variable has reached a preset limit.

audit trail: A record of all the changes made to system parameters and settings, including the date and time of each change as well as the user identification of who made the change.

B

backbone cabling: Network cabling that runs from the entrance facility (where the telecom utility brings in Internet) to and between one or more telecom closets.

bandwidth: The maximum rate at which bits can be transmitted over a certain media type.

baseband: A data transmission method where data is sent without frequency modulation.

bit: A unit of information that has only two states, which are commonly labeled as "on" and "off" or "1" and "0."

building operator: A day-to-day building system manager who is responsible for all HVAC, electrical, plumbing, and life-safety systems.

C

cable tester: A test tool that can check the integrity and proper connector pinouts of a multiconductor cable.

cable tracer: A test tool that can identify a particular cable out of many at anywhere along its length.

client computer: A networked computer that requests information from a server or other nodes.

coaxial cable: Cable that consists of an inner conductor surrounded by an insulating layer, surrounded by a conductive shield, and then a protective covering.

collision: A data transmission overlap that corrupts the data.

crossover cable: A cable that is terminated with some pin-outs reversed on each connector.

crosstalk: The induction of signals onto a conductor from an adjacent conductor.

D

digital multimeter (DMM): An electrical test tool that can measure multiple parameters, commonly voltage, current, and resistance.

Domain Name System (DNS) server: A computer server or router function that translates hostnames into IP addresses so that messages can be routed properly.

dynamic address: An address that can be automatically changed due to network changes.

Dynamic Host Configuration Protocol (DHCP) server: A common router feature that automatically assigns IP addresses to nodes as they are connected to the router and then keeps track of the addresses.

F

field controller: An electronic device that makes control decisions for a specific piece of HVAC equipment based on input information and sends signals as needed to actuators in order to maintain configured setpoints.

field device: An input or output device.

firewall: A routing function that prevents unauthorized access or the transmission of certain message types between the network and the Internet.

H

horizontal cabling: Network cabling that runs from a node or wall jack to the telecom closet.

hostname: A unique node name that is human-readable.

hub: A multiple-port network device that sends all messages received from any port back out to every port.

HVAC controls technician: A technician that installs, configures, maintains, and troubleshoots the equipment and electronics of HVAC system controls.

I

Internet Protocol (IP) addressing: A method of assigning addresses to nodes on a network.

IT technician: An expert in the hardware, software, and configuration of the computer networking infrastructure in a facility.

L

local-area network (LAN): A communication network that spans a relatively small area.

M

media: The cabling that connects the nodes to form the network.

media access control: The method used by nodes to determine if a network is clear of traffic before transmitting data.

media access control (MAC) address: A unique and permanent address that is hard-coded, or permanently saved, into the device by the manufacturer.

N

network: A system of nodes and other devices that are joined using connecting media that provides the means of data communication between the nodes.

network analyzer: A test tool that listens passively to network traffic, recording message packets and saving them for later analysis.

network device: A type of switchboard that allows three or more nodes to communicate with each other.

network interface controller (NIC) card: A card (a circuit board with external ports) installed in a node that allows the node to communicate over a network via a particular type of media.

node: A networked computer (sometimes a very simple one) with a unique network address that sends and/or receives data.

O

Open Systems Interconnection (OSI) model: A way of standardizing the division of a communication system's internal functions by separating it into seven different layers.

optical fiber cable: A cable containing glass or plastic fiber used to transmit information using light.

P

packet: A sequence of bits consisting of a message surrounded by additional information needed to facilitate the transmission and reception of the message.

password scheme: A comprehensive plan for safeguarding user access levels, such as limiting who has access to which inputs, outputs, devices, and functions.

patch: A software update, typically for operating systems, installed to fix bugs and/or improve system security.

patch cable: A relatively short cable used to make connections between two jacks within a telecom closet.

patch panel: A rack-mounted faceplate with many network jacks, each of which terminates a cable run from another part of the building.

port: 1. A physical connector built into a device to allow the connection of a particular type of cable. **2.** A virtual communication connection used by nodes to exchange data directly with certain application programs on other nodes.

protocol: A standard or defined procedure that describes a specific communication network function.

R

router: A network device that manages communication between different sections of a network based on IP address.

S

server: A dedicated computer that provides application services to other computers on a network.

service pack (SP): A package of additional operating system components and updates added to improve system operation.

shielding: Metal foil or braided wire that surrounds wires in order to prevent electromagnetic interference.

static (fixed) address: A permanently assigned address.

subnet masking: An addressing scheme that defines the division between a node's network address and host address.

subnetwork: A subdivision of a larger network that is made up of nodes that communicate frequently.

supervisory controller: An electronic device that contains software-based control functions, coordinates information between field controllers, and may contain high-level control logic.

switch: A multiple-port network device that uses MAC addresses to forward network messages to only the port of the intended recipient.

T

telecom closet: A small room dedicated for housing most network devices and being the collection point of network cabling from a building area.

throughput: The actual rate of bit transmission at a specific moment.

topology: A network arrangement that configures the flow of message traffic.

trend data: A large set of values for system variables, such as for temperature, humidity, pressure, and other variables, that are recorded at regular intervals.

twisted-pair cable: A cable that consists of two or more pairs of wires, with each pair twisted individually.

U

utility: A software tool that helps analyze, maintain, configure, or optimize the operation of a computer.

V

virtual private network (VPN): A software feature that provides an encrypted and secure method of remote access that mimics a local, physical connection to a LAN.

virtual server: A software emulation of a complete server operating system, separate and independent from other virtual servers on the same hardware.

W

wide-area network (WAN): A communication network that spans a relatively large area.

wireless access point (WAP): A network device that is connected to a wired LAN and provides access to the LAN by wireless nodes via radio-frequency signals.

wireless meter: A test tool that detects the presence and type of wireless networks and measures the corresponding signal strengths.

wireless network adapter: A component that transmits and receives radio-frequency signals for a network node.

wireless networking: The use of radio frequencies as a communication medium for computer networks.

Index

Page numbers in italic refer to figures.

A

access blockage, 81–82
access points, 58–59, *59*
addresses, 8
administrators, 3
Advanced Research Projects Agency (ARPA), 5
alarms, 44, *45*
analyzers, 66–67, *67*
antivirus software, 40–41
ARPA (Advanced Research Projects Agency), 5
audit trails, 45

B

backbone cabling, 12, *13*
backing up servers, 52
bandwidth, 7
baseband, 10
binary numbering, *20*, 20
bits, 6–7
browser pop-ups, 82–83, *83*
browsers, 35–36, *36*, 86
building operators, 1

C

cable markings, *78*
cable problems, 75–79, *77*, *78*, *79*
cable terminations, *15*, 15, *16*
cable testers, 65–66, *66*, 76, *77*
cable tracers, 66, *67*, 77, *78*
cabling
  backbone cabling, 12, *13*
  coaxial cables, *12*, 12
  crossover cables, 15, *16*, *78*, 78
  horizontal cabling, 12, *13*, *14*
  optical fiber cables, *12*, 12
  patch cables, *14*, 14–15, 75–77, *77*, *78*, 78
  twisted-pair cables, *10*, 10–12, *11*, *15*, 15, *16*
CAT ratings, 10–11, *11*
client access licenses (CALs), 51
client computers, 33, 75
Cloud computing, *53*, 53
coaxial cables, *12*, 12
collaboration among staff, 4
collisions, 8
command modifiers (switches), 69
command prompt tools, *68*, 68–69
common-bus topologies, *6*, 6
computer-based equipment, *2*, 2
computer configurations, 33
computers. *See* client computers
computer software, 34–38, *35*, *36*, *37*, *38*
configurations, 33
connection speeds, *84*, 84
controller LED indicators, *64*, 64–65
controllers. *See* field controllers; web-based
  supervisory controllers
control networks, 25–28, *26*, *27*, *29*
control panels, 69, 75, *85*, *86*
controls, *2*
controls software, 41
crossover cables, 15, *16*, *78*, 78
crosstalk, 67

D

daisy chain topologies, *6*, 6
data storage and retrieval, 43–45, *44*, *45*
dependencies, 36–37, *37*
DHCP (Dynamic Host Configuration Protocol)
  servers, 18
diagnostic tools. *See* troubleshooting tools
digital multimeters (DMMs), *65*, 65
Domain Name System (DNS) servers, 21
.NET Framework, 37
drivers, 38
dynamic addresses, 18, *19*
Dynamic Host Configuration Protocol (DHCP)
  servers, 18

E

enclosures, supervisory controllers, *31*

Ethernet networks
building and telecom closet cabling, 12–14, *13*, *14*
cable terminations, *15*, 15, *16*
common problems, 53
Ethernet, definition of, 9
Internet Protocol (IP) addressing, 18–21, *19*, *20*
media access control (MAC) addresses, 17–18, *18*
media (cabling) types, 9–12, *10*, *11*, *12*
network devices, 21–24, *22*, *23*
network interface controller (NIC) cards, *16*, 16–17, *63*, 63–64
testing, 65
Ethernet ports, *85*, 85

F

fault LEDs, 64
field controllers, 26, *27*
field devices, 25
firewalls, 24, *40*, 81–82
fixed (static) addresses, 18, *19*
flashing devices, 31

G

gateways, 24. *See also* routers
graphics, 45, *46*

H

handheld testers, *65*, 65–67, *66*, *67*
hardware tools, *63*, 63–67, *64*, *65*, *66*, *67*
hexadecimal numbering, *17*, 17
horizontal cabling, 12, *13*, *14*
hostnames, 21
hosts. *See* nodes
hubs, 21, *22*
HVAC controls technicians, 2
HyperTerminal, 69–70

I

IEEE (Institute of Electrical and Electronics Engineers) standards, 59, *60*
IIS (Internet Information Services), 37
imaging software, 52
installation
of servers, 51–52
of web-based supervisory controllers, 29, *31*, 31
Institute of Electrical and Electronics Engineers (IEEE) standards, 59, *60*
integrated networks, 25, *26*
interfaces, web-based supervisory controllers, 28, *29*
internal (restricted) wireless networks, 55
Internet Information Services (IIS), 37
Internet of Things (IoT), *30*, 30
Internet Protocol (IP) addressing, 18–21, *19*, *20*, 79–80, *80*, 83
ipconfig command, 69
IT closets. *See* telecom closets
IT policies, 33–34
IT technicians, 2–3

J

Java®, 35, *36*, 86–87

L

LANs (local-area networks), 5, *6*
LED indicators, *63*, 63–65, *64*
LEDs (link indicators), *63*, 63
licenses, client access, 51
licenses, software, 38
link indicators (LEDs), *63*, 63
local-area networks (LANs), 5, *6*

M

MAC (media access control) addresses, 17–18, *18*
maintenance of servers, 52
media, 9–10. *See also* cabling
media access control, 8
media access control (MAC) addresses, 17–18, *18*
mesh network topologies, *6*, 6
message data, 6–7, *7*
meters, wireless, 67
Microsoft® Windows® versions, 34, *35*

N

.NET framework, 37
netstat command, 69
network access, 38–39
network administrators, 3
network analyzers, 66–67, *67*, *84*, 84
network architectures, 5, *6*, 25, *26*
network configuration tools, *68*, 68–69
network device LED indicators, 64
network devices, 21–24, *22*, *23*
networking history, 5
networking infrastructure, 2, *3*
networking personnel, 1–4
network interface controller (NIC) cards, *16*, 16–17, *63*, 63–64
network ports, *85*, 85
networks, 5–8, *6*, *7*. *See also* control networks; Ethernet networks; wireless networking
network status command, 69
network topologies, *6*, 6
network utilities, 37
NIC (network interface controller) cards, *16*, 16–17, *63*, 63–64
nodes, 5, *6*, 8

O

Open Systems Interconnection (OSI) model, 8, *9*
operating systems (OSs), 34, *35*, 48–49
optical fiber cables, *12*, 12

P

packets, *7*, 7
passwords, 39–40, 45–46, *58*, 58, 84–85, 86
password schemes, 45–46
patch cables, *14*, 14–15, 75–77, *77*, *78*, 78
patches, 34, 49
patch panels, *13*, 13, *14*
peripherals, *38*, 38
personnel, 1–4
ping command, 68–69
policies, 33–34
pop-up blockers, *36*, 36
pop-ups, 82–83, *83*
ports, 16, 21, *85*, 85
printers, *38*, 38
protocols, 8, *9*
public wireless networks, 55
purchasing servers, 47

R

rack mount server cabinets, *47*, *48*
radio-frequency signals, 58
reimaging devices, 31
remote access, 39
remote access software, 50, *51*
restricted (internal) wireless networks, 55
ring topologies, *6*, 6
RJ-45 connectors, *15*, 15
routers, *23*, 23–24, 78–79, *79*

S

screened shielded twisted-pair (SSTP) cabling, *10*, 10
screened twisted-pair (ScTP) cabling, *10*, 10
security, 39–41, *40*, *58*, 58
separate networks, 25, *26*
serial ports, 69, *70*
serial terminal capture, 69–70, *70*
server rooms. *See* telecom closets
servers
  backup, 52
  capabilities of, 43–46, *44*, *45*, *46*
  definition of, 43
  imaging, 52
  installing (setting up), 51–52
  locations of, *48*, 48
  maintenance of, 52
  operating systems (OSs), 48–49
  physical vs. virtual, 47
  purchasing, 47
  remote access software, 50, *51*
  sizes of, 43, *44*
  software, 48–51, *50*, *51*
  troubleshooting, 52–53, *80*, 80–81
  utilities, 49, *50*
service packs (SPs), 34, 49
shielded twisted-pair (STP) cabling, 10
shielding, 10
single sign-on (SSO) capabilities, 46
64-bit architectures, 49
sniffer units, 66, *67*

software
  antivirus, 40–41
  common problems, 53
  computer, 34–38, *35*, *36*, *37*, *38*
  controls, 41
  licensed, 38
  server, 48–51, *50*, *51*
  troubleshooting, 41
  updates for, 31, 82, 85, 86–87
software tools, *68*, 68–70, *70*
SPs (service packs), 34, 49
SQL (Structured Query Language) software, 37, 49, *50*
SSO (single sign-on) capabilities, 46
SSTP (screened shielded twisted-pair) cabling, *10*, 10
star topologies, *6*, 6
static (fixed) addresses, 18, *19*
STP (shielded twisted-pair) cabling, 10
Structured Query Language (SQL) software, 37, 49, *50*
subnet masking, 19, *20*
subnetworks, 24
supervisory controllers, 26–28. *See also* web-based supervisory controllers
switches, 21–23, *22*, *23*, 78–79
switches (command modifiers), 69
system administrators, 3
system security, 39–41, *40*
system utilities, 37

T

technicians, 2–3
telecom closets, *9*, 13, *48*, 48, 77, *78*
telephone line network, 5
terminal software, 69–70
terminations, cable, *15*, 15, *16*
testers, *65*, 65–67, *66*, *67*
thermostats, wireless residential, *57*, 57
thicknet cables, 12
thinnet cables, 12
third-party network tools, 70
32-bit architectures, 49
throughput, 7
tools
  controller software, 46
  hardware, *63*, 63–67, *64*, *65*, *66*, *67*
  software, *68*, 68–70, *70*
  troubleshooting (diagnostic). *See* troubleshooting tools
topologies, *6*, 6
trace route command, 69
tracert command, 69
tracing cables, 66, *67*
trend data, *44*, 44
troubleshooting
  browser pop-ups, 82–83, *83*
  browsers, 86
  cables, 75–79, *77*, *78*, *79*
  client computers, 75
  common parts of a network, 73, *74*
  common scenarios chart, *76*
  connection speeds, *84*, 84
  Internet Protocol (IP) addresses, 79–80, *80*, 83
  Java®, 86–87
  networks ports, *85*, 85
  passwords, 84–85, 86
  resources for, 74–75
  routers, 78–79, *79*
  servers, 52–53, *80*, 80–81
  skills for, 73
  software, 41
  software updates, 82, 85, 86–87
  switches, 78–79
  tools for. *See* troubleshooting tools
  virtual private networks (VPNs), 82
  web-based supervisory controllers, 32
  wireless networks, 60, *81*, 81
troubleshooting tools
  cable testers, 65–66, *66*, 76, *77*
  cable tracers, 66, *67*, 77, *78*
  command prompts, *68*, 68–69
  digital multimeters (DMMs), *65*, 65
  LED indicators, *63*, 63–65, *64*
  network analyzers, 66–67, *67*, *84*, 84
  serial terminal capture, 69–70, *70*
  third-party network tools, 70
  vendor tools, 70
  Windows® Control Panel, 69
  wireless meters, 67
turnkey servers, 47
twisted-pair cables, *10*, 10–12, *11*, *15*, 15, *16*

U

unshielded twisted-pair (UTP) cabling, *10*, 10

utilities, 37, 49, *50*, 60, *61*

V

vendor tools, 70
vendor utilities, 37
virtual private networks (VPNs), 39, 82
virtual servers, 47

W

WANs (wide-area networks), 5, *6*
WAPs (wireless access points), 58–59, *59*
web-based supervisory controllers
  enclosures, *31*
  installing, 29, *31*, 31
  interfaces for, 28, *29*
  LED indicators, *64*, 64–65
  software updates, 31
  supervisory controllers, definition and examples of, 26–28, *28*
  troubleshooting, 32
  web-based access, 28, *29*
web browsers, 35–36, *36*
wide-area networks (WANs), 5, *6*
Wi-Fi®. *See* wireless networking
Windows® Control Panel, 69, 75, *85*, *86*
Windows® dependencies, *36*, 36
Windows® Installer, 36
wired access, 39
wireless access, 39
wireless access points (WAPs), 58–59, *59*
wireless meters, 67
wireless network adapters, 59
wireless networking
  advantages, *56*, 56
  definition of, 55
  disadvantages, 56, *58*, 58
  public, 55
  residential thermostats, *57*, 57
  restricted (internal), 55
  standards, 59, *60*
  troubleshooting, 60–61, *81*, 81
  utilities, 60, *61*
  wireless access points (WAPs), 58–59, *59*
  wireless and wired networks, *55*
  wireless network adapters, 59
wireless residential thermostats, *57*, 57
Wireshark®, 70